WHALES
OF THE
WORLD

WHALES
OF THE
WORLD

Nigel Bonner

BLANDFORD

Blandford Press
An imprint of Cassell
Artillery House, Artillery Row, London SW1P 1RT

First published 1989
Distributed in Australia by
Capricorn Link (Australia) Pty Ltd
PO Box 665, Lane Cove, NSW 2066

British Library Cataloguing in Publication Data

Bonner, Nigel
    Whales of the world.
    1. Dolphins & whales
    I. Title
    599.5

ISBN 0 7137 1971 0

Typeset by Graphicraft Typesetters Ltd,
Hong Kong
Printed in Yugoslavia by Papirographica

Contents

Acknowledgements   7

Preface   8

1   Whales Great and Small   9

2   The Great Whales – the Rorquals   28

3   Right Whales and Bowheads   48

4   Humpbacks and Grey Whales   68

5   Sperm Whales – Divers in the Deeps   84

6   The Oceanic Dolphins   102

7   Killers and Their Kin   130

8   Narwhals and Belugas   146

9   Beaked Whales and River Dolphins   156

10   Whales and the Modern World   169

Appendix: A Classification of the Order Cetacea   183

Guide to Further Reading   186

Index   188

For Jennifer

and to the memory of
Leonard Harrison Matthews MA, ScD, FRS
1901–1986

Acknowledgements

I could not have written this book without drawing heavily on the work of others. They are too many to name here, but I gratefully acknowledge my debt to them. Many friends and colleagues assisted by providing photographs to use for reference or illustration, not all of which it was possible to use. Knowing how tedious it is to sort out pictures, I am very grateful to them. My wife provided continual support and encouragement while I wrote, and also helped in innumerable other ways.

Finally, I would like to remember an outstanding personality in the select world of cetologists. Leo Harrison Matthews preceded me in South Georgia by some 30 years. He, too, found much to fascinate him there in the whales and the whalers. He went on to become a foremost authority on whales, and certainly one of their most entertaining chroniclers. His learning and good company were freely available to all who sought them, and he provided me with much encouragement during my time in South Georgia and subsequently. I hope he would have approved of this book.

Illustration credits

Colour

BAS: p. 158
Michael Bryden: p. 119
Bill Doidge: p. 151 (top)
D.R. Gipps: p. 34
Christina Lockyer: p. 110
Tony Martin: pp. 23, 70, 127, 131 (bottom), 142, 151 (bottom)
William E. Perrin: pp. 82, 99, 159, 175

Black and white

BAS: pp. 35, 36, 132, 134, 158
Margaret Barstow: p. 167
T. Dicks, courtesy Mrs F. Hayes and G.J.B. Ross: p. 67
Peter Evans: pp. 121, 126
L. Harrison Matthews, courtesy Mrs D. Harrison Matthews: pp. 162, 172
Tim Waters: pp. 33, 85, 125, 145

All other photographs were taken by the author, who also drew the line illustrations.

Note: Two spellings of 'ton' have been used in this book: 'ton' (2240 lbs) refers to weights taken from historical sources, and 'tonne' refers to the metric tonne (1000 kg).

Preface

Whales have had a fascination for me since I was a boy. This was reinforced when I went to South Georgia in 1953, at the peak of the Antarctic whaling industry, and spent nine whaling seasons there. My feelings as a conservationist were outraged at the sight of so much carnage and the evident indications of the impending doom of the whales and of whaling, although, at the same time, it was impossible not to feel a sympathy for the whalers – fine seamen, who had been brought up in communities that knew no other trade. By the time the world woke up to the plight of the whales, whaling had already collapsed for economic reasons – there were insufficient whales left to support the industry! Some small-scale whaling continued, and, indeed, continues still, but in the western world people had begun to realise that whales, in all their diversity, were part of the world ecosystem and deserved their place on this planet.

Much of this public sympathy derived from a greater familiarity with whales, largely brought about by the exhibition of small whales, particularly bottlenose dolphins, in animal parks and oceanaria. Seeing so much of whales, people wished to learn more about them, and in doing this a considerable empathy developed between humankind and these extraordinarily aberrant mammals, the whales.

I hope this book will do a little more to satisfy people's curiosity about whales. Perhaps it will even stimulate some readers to go on to learn more about these strange and wonderful creatures. This is a small book, so I have had to be selective about what I put into it, and an inevitable consequence of this is that I will have treated some topics rather scantily and left out others altogether. I hope that those wanting to learn more about whales will have the patience to refer to some of the excellent works listed in the section on Further Reading.

Nigel Bonner

Chapter 1
Whales Great and Small

Inevitably the word 'whale' carries the connotation of immense size and, indeed, the whales do include the greatest animals ever to have inhabited this Earth. But although all whales are large, as mammals go, not all of them are huge. The smaller species, the dolphins and porpoises, are not generally regarded as whales, but are equally members of the order Cetacea, and any survey of whales as a group would be incomplete without consideration of these lesser species as well as giants like the sperm whale or the mightiest of them all, the blue whale. I shall refer to all the members of the order as 'whales'.

What is it that unites these very different marine mammals into one recognisable group, for no group of mammals, except the bats, is so instantly recognisable? The answer is that whales live in water and this has modified their appearance and behaviour until they are far removed from what we think of as a 'typical' mammal.

By far the greatest majority of vertebrates that live in water are fish. Whales and fish have something of a common appearance; indeed, so close was this resemblance that for many years whales were classed as fish (though Aristotle knew better 24 centuries ago). Both fish and whales, and the extinct aquatic reptiles of the Cretaceous period, the ichthyosaurs and the plesiosaurs, adopted a similar bodyform because they had to cope with the special properties of water (Fig. 1.1).

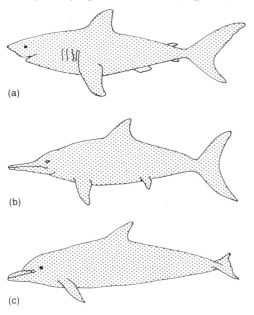

(a)

(b)

(c)

Fig. 1.1 Outlines of (a) a fish (a shark), (b) an extinct reptile (an ichthyosaur) and (c) a modern cetacean (a dolphin), to show the similarity of bodyform brought about by the marine environment.

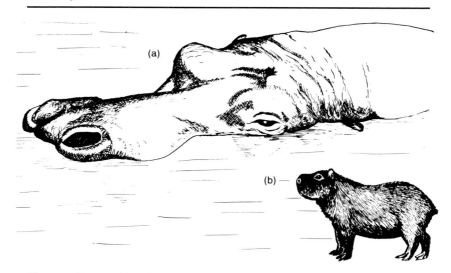

Fig. 1.2   Two moderately aquatic mammals, the hippopotamus (a) and the capybara (b). The eyes and nostrils are high up on the head, allowing the animals to breathe and see while almost completely submerged.

Water is the medium in which vertebrates evolved, but the primitive mammals were small shrew-like terrestrial creatures, which scurried about the undergrowth feeding, probably at night, on invertebrates. For them, the ability to move quickly over the ground, so as to escape predators, and a good nose and set of whiskers, so as to find their prey, were important factors. Later, as the primitive mammals grew larger and began to exploit the open country, they became more diurnal and good eyes became another important factor.

Whales returned to the water secondarily. We do not know the reason for this, but it was most likely associated with the search for fresh food resources to exploit. Many mammals have become secondarily aquatic to varying degrees in the course of evolution. The hippopotamus is a well-known example from Africa, and the capybara a less familiar one from South America. Both of these are really quite ordinary mammals. They have some modifications for an aquatic life – eyes and nostrils high up on the profile of their heads, for example (Fig. 1.2), but both still retain strong links with the land.

Seals and sea lions have gone a good deal further down the road to an aquatic existence. Their body shape is noticeably different from that of other mammals (Fig. 1.3). Their limbs have been withdrawn into their bodies (the armpit and crutch in a seal are at the level of the wrist and ankle respectively) and modified to an extent which makes them clumsy movers on land. Seals and their kin spend the greater part of their lives in the water, but all must return to land (or on to solid ice) to produce their young.

Only two groups of mammals have managed to separate themselves completely from ties with terra firma. These are the Sirenia, or sea cows, (Fig. 1.3), and the whales. The Sirenia are an odd group. Never very

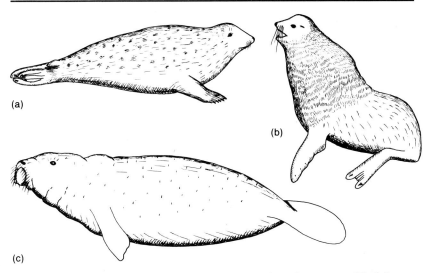

Fig. 1.3   Besides whales, other mammalian groups have become modified for an aquatic existence: (a) a true seal, (b) a fur seal and (c) a manatee, or sea cow.

abundant since they first evolved, there exist today only four species – three manatees, one from the West Indies and Florida, one from the Amazon, and one from West Africa – and the dugong, which ranges widely over the Indian Ocean and the Western Pacific in coastal waters. Two hundred years ago there was a fifth species – Steller's sea cow. This, by far the largest of the group, reaching about 8 m (26 ft 3 in) in length, and weighing perhaps 6 tonnes, was discovered in 1741 by a party of shipwrecked Russian explorers in the sub-Arctic Pacific. Less than 30 years later these huge harmless creatures were extinct.

None of our surviving sirenians has a flourishing population; perhaps the dugong is best off, although its slow reproductive rate, a feature typical of the sirenians, makes the dugong vulnerable to sudden catastrophes. Sirenians represent one of the backwaters of evolution – a group which evolved to take advantage of a particular resource (abundant sea-grass meadows in shallow tropical seas), but which have failed, in the succeeding 50 million years, to adapt to changing conditions and diversify their way of life so as to seize opportunities as they arose.

What a contrast the whales present! These too have cast off all their associations with the land, but the whales, great and small, have embraced the seas from the Equator to the Poles, and from muddy estuarine waters to the deep blue ocean. There is variety of form (Fig. 1.4) and behaviour, and some of the cetaceans are still among the most abundant mammals on Earth. (Species like the common dolphin or the spotted dolphin are numbered in millions.) In terms of biomass, that is, the total mass (or weight, as we loosely say) of the population alive at one time, the great whales, like the fin and the blue, although never very numerous in an absolute sense, were once also highly abundant species.

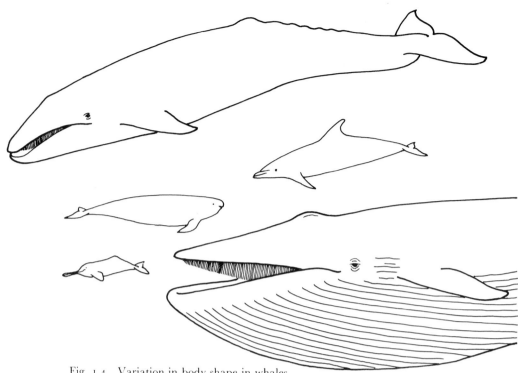

Fig. 1.4   Variation in body shape in whales.

Life in water

But let me return to the role of water in modelling the way of life of the whales. What are the characteristics of water that have so greatly modified the whales from their terrestrial ancestors? Firstly, it is important to realise that we are not considering the differences between water and dry land, but between water and air, for a terrestrial mammal lives in air, with only a minor contact with the substrate (a very important contact, but one that affects less of the general body plan than the air surrounding it). Both air and water are fluids, but water is a very much denser fluid than air and a great deal more viscous.

Archimedes told us that a body displaces its own volume of fluid. We tend to think of this as a body weighing less in water than it would in air. But air, too, has a density (about 1.22 kg per cubic metre, or 0.076 lb per cubic ft) so a body in air weighs fractionally less than it would in a vacuum. Suspend that same body in water, or, better still, in sea water, which has a density of about 1,025 kg per cubic metre (64 lb per cubic ft), and it will weigh a great deal less. If that body happens to be a mammal, small or large, the chances are that it will weigh practically nothing, for the mammalian body has a specific gravity very close to that of water. A whale in the sea is, then, in practical terms, weightless. This at once changes the selective pressures on its bodyform.

For a mammal to live on land, it requires a skeleton to support it. Consider a horse, for example. Its trunk, containing its viscera, approximates to a cylinder suspended from a jointed bony girder, the vertebral column (Fig. 1.5). The head (which needs considerable mobility, so that the horse can lower its mouth to the ground to feed) is attached to the trunk by a stout muscular neck. This needs to be muscular, for the head is heavy and is supported by a system of cantilevers formed by the muscles running from the dorsal spines of the vertebrae of the neck and chest to the back of the skull.

The trunk and head are supported clear of the ground by the legs. These are jointed compression struts, for they have not only to carry the weight of the body, but also to withstand considerable shock when the horse moves over the ground and lands on its feet at the end of a stride. It is the need for movement that has dictated the form of the legs. Horses, of course, are highly specialised movers and their legs are not very typical, but such differences are not important in this comparison. When a horse moves, it does so by exerting a force on the ground through the extremities of its legs, the hooves. The thrust against the ground is generated by muscles, chiefly in the shoulder and hindquarter regions, and the reaction against this thrust drives the body of the horse forward. A horse has several gaits, but all operate in this way, the differences involving no more than the order in which the thrust is applied, the magnitude of effort involved and the extent to which longitudinal muscles along the vertebral column are used.

Things are quite different for whales. Floating virtually weightless in the water, a whale's skeleton has little to do in the way of supporting the body. This is reflected in the nature of the bone that composes it. French prisoners of war in Napoleonic times made a little money, while in prison on Dartmoor, by carving beautiful and intricate ship models from the beef bones that came with their daily rations. The hard, ivory-like bone was an ideal material for such delicate work. Had the prisoners been fed on whale meat, they would have had to find a different way of earning their pennies. Whalebone (except in some specialised parts) is soft and porous and suited only to crude, lumpy carvings, although some of those done by aboriginal peoples of the Arctic are objects of great beauty.

The body skeleton of a whale (Fig. 1.5) consists of little more than a jointed rod down the middle, which provides some longitudinal support and attachment sites for muscles. The individual bones that make up the

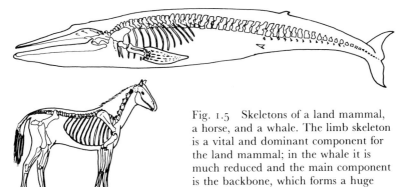

Fig. 1.5   Skeletons of a land mammal, a horse, and a whale. The limb skeleton is a vital and dominant component for the land mammal; in the whale it is much reduced and the main component is the backbone, which forms a huge compression strut.

13

vertebral column are much simpler in the whale than in the horse, and we might rightly suppose from this that the range of movement in the whale is far less than in the horse, and enormously less than in a more limber animal such as a cat. A whale has little in the way of a neck. It has no need to lower its head to gather food from the ground in front of it, nor to raise its head to browse above. In consequence, we find that whales have very short necks in relation to their total size, and these necks have little mobility, the seven cervical vertebrae that make up the neck in all mammals (except some sloths) being reduced or even fused together in the whole.

With no need for support against gravity, the whale, equally, has no need for the four strut-like limbs found in the horse. Whales, in fact, have dispensed with hindlimbs altogether and modified the forelimbs to simple paddles. In the absence of a firm substrate against which to exert force, typical mammalian limbs have an inefficient locomotory function. If one watches a dog paddling in the water, one can see how ineffective is the application of a limb designed for terrestrial locomotion when it comes to progressing in water, while dogs, like horses, one very good at running on land.

Water is a fluid, hence it yields to a force applied to it. To obtain a useful reaction, therefore, it is necessary to apply the force over a wide area. Whales have developed a completely new organ for this purpose. This is the tail, or rather the expanded flukes that have developed on either side of the tail. The flukes of a whale form what is known to anatomists as a neomorph, that is to say, they are a new structure that has appeared in the course of evolution, rather than a modification of an existing structure. They consist mainly of fibrous tissue (I shall describe their structure in greater detail when I deal with swimming in Chapter 6) and are quite different from the tail of a fish, which is supported by a bony framework.

The whale swims through the water by beating its tail up and down, the power for this coming from massive blocks of muscle running along the axis formed by the vertebral column. Once a whale has got underway and begun to move (and a lot of energy must be expended in accelerating a 100-tonne blue whale from rest to even a modest speed), the main resistance it encounters is that provided by the viscosity of the water and its friction against the whale's skin. For a terrestrial mammal the resistance of the air through which it moves is a very minor matter, even when travelling at top speed. But because the viscosity of water is so much greater, this resistance becomes a force to be reckoned with, in evolutionary terms, for whales, even when moving quite slowly.

Streamlining

The response of whales to this force, throughout evolutionary history, has been to streamline their body shape. Streamlining is a rather vague term when applied to living animals, but basically it implies possession of an elongated body, tapered at both ends, but with the forward end blunter than the hind end. Fish and whales had discovered the secret of the bulbous bow long before ship designers!

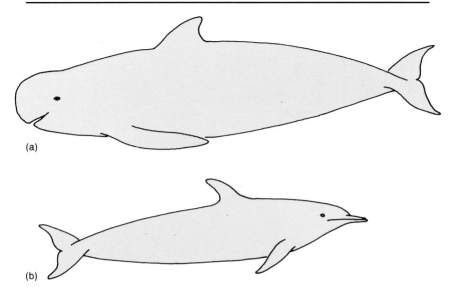

Fig. 1.6   Pilot whale (a) and common dolphin (b). Although head shapes differ, both are efficient swimmers.

Whales show a great variety of different headshapes. All are rounded to some extent, and it may be that the different patterns are related to different modes of swimming. A pilot whale, with its great bulging forehead, or pot (it is really the animal's upper lip!), looks very different in profile from the slender-beaked outline of the common dolphin (Fig. 1.6). Yet, both progress efficiently through the water, although it is clear from even casual observation that they swim in different ways.

Streamlining implies not only the adoption of a basic body shape, but also the elimination of as many as possible of the structures that would interfere with the even flow of water over the body's surface. As we have seen, the whales have dispensed with their hindlimbs in favour of a newly developed locomotory organ, the flukes. Forelimbs are retained, but in the form of paddles. These are generally rather small in comparison with the size of the whale, and their musculature and attachment to the body indicate that they have no function in generating thrust for locomotion. The flippers of a whale are used as vanes, whose angle of attack, as the creature moves through the water, can be varied so as to make small alterations in course possible. A similar function can be seen in the manoeuvring vanes fitted to small underwater vehicles.

Other protrusions common on the body surface of most mammals have been eliminated in whales. A whale has no trace of external ears, and the male genitals are carried internally. The penis, of course, is extruded when males are sexually excited, and some male whales, at least in captivity, have the habit of extruding the penis to use it as a sensitive organ of touch. In the female the breasts are internal, and even when she is producing milk the outline of her body remains smooth and sleek. The nipples are contained within narrow slits on either side of the vent, and remain hidden except when stimulated by the sucking calf. This makes it a difficult matter to tell the sex of whales in their natural environment, although in stranded whales the sex is often grossly apparent. Then gases of putrefaction bloat the corpse and the resulting internal pressure may forcibly evert the penis.

An early print of a sperm whale, which stranded on the Dutch coast in the seventeenth century, showed just this condition. Like a good many illustrations of whales, this woodcut was extensively copied, right down to the nineteeth century, and it is interesting to see how a more prudish generation changed the location of the stranding from a sandy beach to a rocky shore, with a large rock conveniently shielding the offending organ from the delicate gaze of the viewer (see page 18).

Not only have the whales eliminated unnecessary protrusions which might hinder their passage through the water, they have also developed an astonishingly smooth skin. Anyone who has had the opportunity to touch a whale in one of the oceanaria (like Sea World in San Diego where the chance to come into physical contact with cetaceans is actively encouraged), will have been fascinated not only by the sense of immediate contact with these strange and fascinating animals, but also by their amazingly smooth skin. No other mammal has so sleek a covering. There are no foldlines or wrinkles, no pores, nor any of the roughness that characterises bare skin areas in other mammals. No fairy princess ever had a bosom so smooth as the back of a pilot whale!

The reason for this is that whales are the least hairy of all the mammals. However, hair is a diagnostic feature of the class Mammalia, and whales have not been so rash as to abandon this important label altogether! A few specialised hairs (or at any rate, their follicles) are found in all whales, and in many these have an important sensory function. Over the general body surface, however, the hair covering has disappeared. And not only the hairs. The follicles have gone also, and so, too, have the associated sebaceous glands.

Hair has been lost in whales because of the frictional resistance that hair shafts would create when the whale moves through the water. In the absence of hair, water flow can take place smoothly over the slippery skin surface, following the streamlines of its outline.

Hairlessness in aquatic mammals is an interesting subject. Whales and sirenians are hairless, but seals and sea lions most definitely are not. Fur seals, of course, retain their hair as a warm covering, which traps a layer of insulating air even when the creature is swimming in cold water. This role is less important in true seals and sea lions, where the insulating role is largely taken over by deposits of fat beneath the skin, but, nevertheless, the hair performs an important function.

The shafts of the hair act as reinforcing rods in the skin and add greatly to its resistance to abrasion. We can see this very clearly in our own species. How many of us have stood up suddenly beneath an open cupboard door and received a most painful crack on the head? But if we are young and have a good thatch, the result is more likely to be a painful bruise rather than a cut. The thatch of hairs and their shafts, penetrating several millimetres into the scalp, provide great mechanical strength against sudden contacts with sharp edges. If one is unlucky enough to be old, male and bald, the consequences of such an accident may be a little more severe and blood may flow. (The scalps of bald men are not, in fact, hairless, but the hair follicles have reverted to producing a very fine infantile down, the shafts of which are not thick enough to provide any useful protection.)

We can see now why seals retain their hair covering. Coming in frequent contact with rocky shores, as they do when they haul themselves out to bask or to breed, seals are regularly exposed to the risk of abrasion, from which they are largely protected by their hair. Whales, being oceanic, shun the shore and are unlikely to come in contact with sharp objects, so the protection of hair is unnecessary for them.

I might add here the comment that consideration of this role of the hair will show how ill founded is the pseudo-scientific hypothesis that attempts to account for human hairlessness by postulating an aquatic phase in human evolution. As I have pointed out, hairlessness is not a specially aquatic feature, being found in only two groups of aquatic mammals. Had our human ancestors been associated with a seashore or lakeside phase in their evolution (and this is very possible), it is not likely that they would have lost their hair as a consequence. With typical hominoid limbs, our ancestors could scarcely have been fast swimmers, so the minor advantage gained by dispensing with hair (on a very poorly streamlined body) would have been far outweighed by the increased risk of damage as a result of contact with rocks. We must seek another explanation for human hairlessness.

Heat retention

In mentioning the role of hair in fur seals, I referred to the fact that it acted as an insulating layer and that this function was served in true seals and sea lions by a layer of fat. If one lives in water, such insulation is very necessary, for another of the properties of water, that differs from those of air, is its capacity to absorb heat from warmer bodies. Mammals are warm-blooded and, except in the warmest parts of the tropics, they are always liable to lose heat to their colder surroundings. Heat lost in this way represents a waste of energy which has to be made up by increased food intake, so mammals adopt various strategies to reduce this loss to tolerable levels.

For most mammals a thick fur covering helps to trap a layer of warm air next to the body and thus, because air is a very poor conductor of heat, heat losses from the body are reduced. In more extreme conditions, the mammal may adopt behavioural patterns which lead it to seek shelter in protected places, or to build a nest where it can keep warm. Some small mammals, in areas where winters are regularly cold and where food supplies may be limited at that season, give up the struggle against cold, go into a torpor and allow their body temperature to drop to only a few degress above freezing. On the return of more favourable conditions, these hibernating mammals wake up and resume a normal active existence.

Mammals which live in the sea have a more acute problem. Heat transfer

An early engraving of a whale. Although labelled 'Mysticetus' (or right whale), this is clearly a sperm whale as the presence of teeth in the lower jaw, but not the upper, shows.

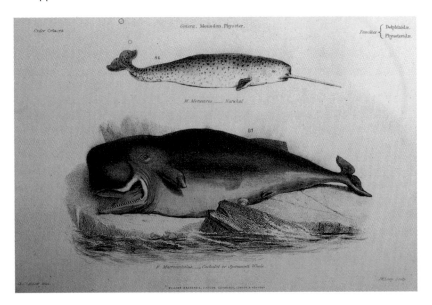

A Victorian version of the same whale as shown *above*. A strategically placed rock has been introduced for modesty's sake.

from a warm body is about 27 times faster in water than in air, for not only does water have a much greater heat capacity (specific heat) than air, it is also a much better conductor of heat. The body temperature of a whale (which is in the range 36–37°C [96.8–98.6°F], about the same as that of humans) is always higher, and usually much higher, than that of the water in which it swims, so the whales are faced with something of a metabolic problem.

Whales have solved this problem in two ways. Body heat is lost in breathing when warm air is exhaled, and some is lost with the faeces and urine, but by far the greatest part of the loss of heat in any mammal is from the body's surface. To reduce such losses, it is possible to reduce the amount of surface or to insulate what surface there is, so that heat is lost more slowly across it. Whales have done both these things.

We have already seen how the whale's body outline has been smoothed by the elimination of projecting appendages in order to reduce drag through the water. The elimination of these surfaces equally reduces heat loss. However, there remains an irreducible minimum after all the non-essentials have gone. Even then, there is a way of reducing this in proportionate terms. It is a simple geometric property that for solids of the same shape, the larger they are the smaller is the relative proportion of their surface to volume.

For most people it is easiest to visualise this in terms of a series of cubes (Fig. 1.7). A cube with sides 1 cm in length has a volume of 1 cubic cm, a surface of 6 sq cm and a surface-to-volume ratio of 6. Double the dimension and the volume becomes 8 cubic cm (2 × 2 × 2), while the surface is 24 sq cm (2^2 × 6); a ratio of 3. Treble it and the volume is 27 cube cm, while the surface is 54 sq cm; a ratio of only 2. And so on.

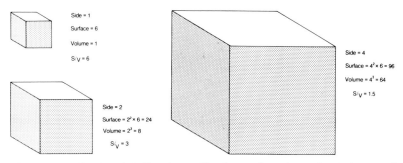

Fig. 1.7   For bodies of similar shape, like these cubes, the surface/volume ratio decreases as absolute size increases.

Whales are not cubes, of course, but the same relationship applies for any set of similar shapes, hence the larger a whale is, the less relative surface it will have and the less heat it will lose across it.

A consequence of this is that whales, even the smallest of them, are all *large* mammals in the literal sense of the word. The smallest cetaceans, creatures like the little La Plata dolphin, or franciscana, which I can scarcely bring myself to call a whale, weigh in the region of 30–50 kg

(66–110 lb). There are few orders of mammals that do not have, as their smallest members, animals much smaller than this. Think of the tiny shrews, bats, weasels and mice. Even in orders that we think of as being generally large mammals – the hoofed animals or artiodactyls, for example – we find minute forms like duikers. Only in the elephants, sea cows and whales do we find no small representatives at all, and it is notable that two out of these three orders are aquatic. (Seals and sea lions are also aquatic and also large mammals, but they are a specialised part of the order Carnivora, which contains some very small forms.)

Whales insulate their bodies by investing them with a thick layer of fat – the blubber. Fat is a good insulator and some of the whales that live in the coldest waters have a great deal of it. In the Greenland right whale, or bowhead, the blubber may be up to 50 cm (20 in) thick! This is exceptional, however; a blue whale might have about 12 cm (4.7 in) of blubber and some of the smaller dolphins no more than 5 cm (2 in). There is much variation in the thickness of the blubber among the different species, or even among different individuals of the same species. Furthermore, the thickness of the blubber varies seasonally. Blubber, in fact, has more than one function. Not only is it an insulator, it is an important food store as well, as we shall see when I come to describe the seasonal feeding migrations of the rorquals in the next chapter.

Blubber forms a passive insulation around the body of the whale, but the

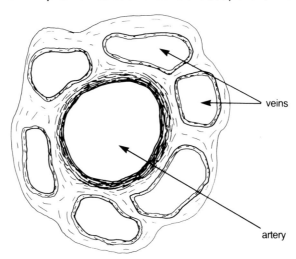

veins

artery

Fig. 1.8   Small veins surrounding an artery in the skin of a whale (a periarterial venous plexus) act as heat-exchangers.

whale also has active ways of retaining its body heat. Parts of the whale's body are devoid of a blubber covering – the flukes, flippers and much of the surface of the head. In these regions the arteries running to the skin are surrounded by complicated spirals of veins (Fig. 1.8). These structures, known by the ponderous name of periarterial venous plexuses, act as counter-current heat-exchangers.

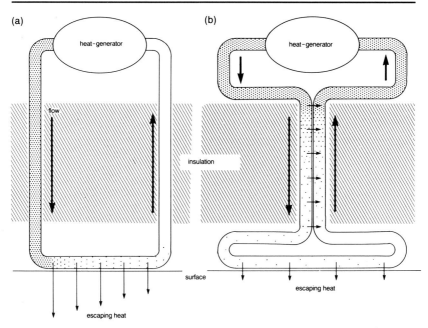

Fig. 1.9   Counter-current heat-exchange. In the simple system on the left (a), heat ducted through the insulation is lost to the exterior; in the counter-current system (b), the warm out-going current gives up its heat to the cold return flow, so that little heat remains to be lost to the exterior.

Counter-current heat-exchangers are devices of considerable importance in engineering, but nature discovered them long before engineers did. The system is easiest to understand by visualising a heat-generator, a boiler for example, within an insulating barrier, with pipes carrying a heated liquid to the cold side of the barrier and back again. In this simple form (Fig. 1.9a) heat will be lost outside the insulation and the returning liquid will be cold. But if, in passing through the insulating layer, the flow and return pipes are in close contact (Fig. 1.9b), heat will be exchanged there, and a cooler liquid will circulate outside the insulation, which will give up less heat because there will be a smaller temperature difference between the contents of the pipe and the exterior. Because the liquid in the two adjacent pipes is flowing in opposite directions (*counter-current*), the warmest part of the flow will meet the warmest part of the return, and *vice versa*, so that the best possible gradient is maintained to exchange the heat.

   In the whale the flow pipe is represented by the the superficial arteries of the skin of the flukes and flippers, and the return pipe by the network of veins surrounding the arteries. These complex branching networks of blood vessels are known as retia (singular rete), from the Latin for a net. (We shall meet them elsewhere in the whale.) The action of these heat-exchangers is controlled by an ingenious mechanism. It is not always in a whale's interest to retain heat. If the animal is very active, the heat produced by muscular activity has to be lost somehow, or the animal's internal temperature would uch a case the arterial pressure increases, causing the artery wall to ⁀ɪ and restrict the return flow of blood through the surrounding veins. The blood is thus forced to return by another set of vessels not close to the arteries, and hence not benefiting from heat-exchange, and thus heat is lost from the system.

Skin, of course, is a living tissue and it is necessary for there to be a blood supply to the outermost layer of the whale's skin to allow for normal growth and repair. However, the presence of an ordinary skin circulation would allow too much heat to be lost. We find that the blubber is traversed by numerous fine arteries, which break up to form a capillary bed beneath the epidermis but outside the insulating layer. These capillaries re-form into veins and pass back into the general circulatory system. Further, besides the capillary bed which is characteristic of most mammalian skin, the whale has a system of vessels of comparatively wide bore, linking the arteries and veins. These are known as arterio-venous anastomoses and are well provided with smooth muscle coats. It seems likely that they are under the control of the animal's autonomic nervous system, which, among other functions, controls the body temperature. If the muscles are relaxed, most of the blood will pass back through the anastomoses, but if the muscles contract, the blood will be forced to move through the capillary bed, making possible repair or regrowth of the skin, or perhaps just dumping heat to the exterior if the whale is overheating.

The ancestors of whales

Mammals are basically a terrestrial group and it is to land-dwelling forms that we must look for the ancestors of the whales. Like that of many other mammalian orders, the origin of the Cetacea is far from clear, but some recent discoveries of early fossil material have thrown light on their ancestors.

About 50 million years ago, during the Paleocene age, there lived a group of mammals, the condylarths (otherwise known as the creodonts), which might be regarded as primitive ungulates. Some of these, the mesonychids, seem to have lived mostly near estuaries and lagoons, and may have been evolving towards a dependence on food from the water. They were bulky mammals, with a varied dentition that indicated that while some were basically herbivorous, others ate flesh. Perhaps they were expanding into the ecological niches left vacant by the mysterious disappearance of the big aquatic reptiles – the ichthyosaurs and plesiosaurs – that had occurred at the end of the Cretaceous period.

The mesonychids were not whales, nor even closely related to them, but the recent discovery of some fossils from Pakistan, appropriately named *Pakicetus*, has provided a link between the most primitive whales, the archaeocetes, and the mesonychids on account of similarities in the teeth, features regarded as of great significance by palaeontologists. This link with the mesonychids, and the rest of the condylarths, makes the whales distant relatives of the artiodactyls (pigs, camels, deer, cattle and the like), as these are clearly related to the condylarths. This evidence from the fossil record is

A humpback whale surfaces to breathe. The wide-open nostrils, or blowholes, are at the top of the head. The lumps on the snout, which are characteristic of humpbacks, are each surmounted by a single sensory hair.

supported by what we know of similarities in blood composition, foetal blood sugar, chromosomes, insulin, uterine morphology and tooth enamel structure from living representatives of both groups.

We know rather little about the appearance or way of life of the archaeocetes, as fossil remains are not very abundant or complete. However, they were elongate aquatic mammals, of moderate size, that is to say, small in comparison with present-day whales. The earliest forms, like *Pakicetus*, for example, probably had the full mammalian complement of four limbs. They would fit very well the usual description of sea-serpents or lake monsters, so perhaps the indefatigable searchers of Loch Ness will yet provide us with an archaeocete. The first archaeocete fossils discovered came from Alabama and Louisiana in the USA and were believed, when they were found in 1832, to be the bones of a large reptile. Hence they were named *Basilosaurus*, or king-lizard.

Basilosaurus owed its great length to an increase in the number of vertebrae behind its thorax, and to an extension in the length of the body of these vertebrae. The processes of the vertebrae are not particularly well developed, and it seems likely that *Basilosaurus* progressed by a rather eel-like wriggling motion, passing S-shaped waves down its body. It probably

23

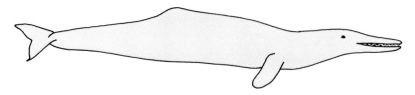

Fig. 1.10 *Basilosaurus*, an early archaeocete, probably looked something like this.

had little or nothing in the way of tail flukes (Fig. 1.10). Although the post-cranial skeleton of the earliest known archaeocetes is already highly adapted to life in the water, the skull remains relatively primitive. The teeth did not exceed the basic number of 44, characteristic of most mammals, and, in contrast to modern, toothed whales, showed some division of function, the teeth at the front of the jaw being simple pegs suited to grasping, while those further back are elongated in the plane of the jaw and have a multi-cusped cutting edge. We can visualise *Basilosaurus* grabbing a fish with its front teeth, and then chopping it into swallowable-sized pieces with the teeth at the back.

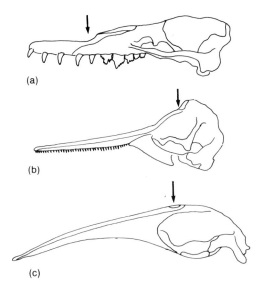

Fig. 1.11 Skulls of an (a) archaeocete and (b) modern toothed and (c) baleen whales. The arrow indicates the position of the nostrils, which have migrated backwards, to the top of the head.

The bones of the skull show none of the telescoping characteristic of modern whales, and the nostrils open about the middle of the snout, so that the 'blowhole', if one can call it such, had migrated only about half way to the posterior position it occupies in most modern whales (Fig. 1.11).

Besides the elongated archaeocetes like *Basilosaurus*, which became

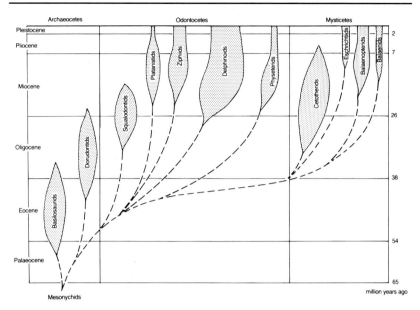

Fig. 1.12 A possible scheme for the evolution of whales.

extinct about the end of the Eocene epoch, or a little later, there were other forms, such as *Dorudon*, with a bodyform more like modern whales. These persisted through the Oligocene period to the beginning of the Miocene, about 25 million years ago. It was during the Oligocene period (between 38 and 25 million years ago) that families of primitive whales, which have clear affinities with the existing suborders, appeared. These were the Agorophiidae and Squalodontidae, which are primitive odontocetes, or toothed whales, and the Aetiocetidae and Cetotheridae, which are primitive mysticetes, or baleen whales.

The squalodonts were a reasonably successful group in the Oligocene era, but had disappeared by the middle of the Miocene. Their teeth were primitive by modern odontocete standards, being clearly divided into incisors, canines and cheek teeth, but they showed many advanced adaptive features for a marine life. The telescoping of the skull was complete and the blowhole had reached its final posterior position on the top of the skull, above and behind the eyes. These changes in the skull may have been associated with the development of acoustic scanning as a means of locating prey or generally detecting underwater features. Squalodonts were not large whales, and probably lived in much the same way as pilot whales or killer whales do today.

It was during the Miocene epoch that the majority of the existing families of toothed whales appeared. Few, if any, of these can be linked with the squalodonts, although the modern river dolphins (platanistids) share a common character in that the temporal opening in the skull is not roofed over with bone, as in other modern odontocetes. Perhaps the modern families, like the squalodonts themselves, are offshoots of the other

25

Oligocene toothed whale group, the Agorophiidae. Squalodonts disappear from the fossil record in the late Miocene era, and agorophiids at the beginning of that epoch.

Baleen whales are so different in many respects from odontocetes that it was once believed that they had a quite separate origin – indeed, some cetologists think so still. However, there are clues which show their common origin. The first time that I dissected a fin whale foetus, lying on the flensing plan by the carcase of its huge mother, I noticed, buried just beneath the gum of the upper jaw, from where the fringe of baleen would have hung had this foetus lived, a row of perfect primordial tooth germs of typical odontocete character. More recently, laboratory studies of the number and shape of the chromosomes, and even of the structure of the DNA and RNA molecules within the chromosomes, have shown that baleen whales are clearly close relatives of the odontocetes, and that the order Cetacea has a real biological identity.

But let me return to the two fossil families, the aetiocetids and the cetotherids, that first appeared in the Oligocene epoch. The aetocetids had differentiated teeth, and so were clearly not yet baleen whales, but they did have a pattern of air sinuses in their skulls, similar to that of present-day rorquals.

The most primitive of the true baleen whales are the cetotherids. These were relatively small whales, often less than 7.5 m (24 ft 7 in) in length. In many ways they resembled the modern grey whale, which is itself the sole survivor of a very ancient line. Cetotherids were abundant through the Miocene era (a great age for whales: what we see today is a mere fraction of what flourished then), but finally disappeared in the Pliocene, being replaced by the modern baleen whales, the rorquals and right whales.

It is tempting to ask the question: why did the whales take to the sea and develop into the forms we see today, and how were these changes brought about? The first part of the question is not really meaningful. Perhaps the best answer we can make is that there were resources in the sea awaiting an exploiter, and the descendants of the mesonychids were better able to compete for the fish stocks than existing predators such as sharks and the remaining few marine reptiles (mostly the saltwater crocodiles). This does not imply that the mesonychids saw a harvest waiting in the sea and became aquatic so as to benefit from it. Rather it is that, of the myriad chance variations that occur as the genes of a species' progeny are resorted or mutate, those that confer an advantage are the ones that tend to survive and hence have the opportunity to leave their token in the fossil record.

This, too, helps to answer the second part of the question. The preservation of small advantageous changes through many generations will eventually create major changes in an animal's structure and behaviour. For some structures it is comparatively easy to visualise how such evolution might come about. The development of the tail flukes, a completely new structure in the mammalian series, with their associated muscles and enlarged vertebral processes, could well have come about in imperceptible stages, each slight development conferring a small, but significant, advantage in swimming.

It is far less easy to see how a structure such as the baleen filter bed of the mysticetes could have developed gradually. Baleen is made of keratin, the horny substance that protects the skin of mammals (and most other vertebrates) from the environment. Many mammals, including humans, but more conspicuously the dog, have keratin ridges on the roof of the mouth. It is believed that baleen developed from such palatal ridges, though it is difficult to see what advantage would be conferred by the initial stages in the development of the filter bed. Nevertheless, our inability to imagine what these advantages were does not mean that they did not exist. With more knowledge of the way of life of the ancestors of modern baleen whales, we might be able to ascribe a function (sieving bottom-dwelling invertebrates out of the mud?) to even the shortest fringe of baleen.

Sometimes evolution takes a backward step that surprises us. Although no modern whale normally has any external hind limbs, it does happen that very occasionally a whale is found in which more extensive remnants of the hindlimbs exist. A humpback has been taken with hindlegs that stuck out more than a metre from its belly, and which contained separate bones and cartilage corresponding to the usual limb segments. Victor Scheffer, an eminent American marine mammalogist, commented: 'How vanishingly small are the odds of a gene persisting through millions of years, inactive, yet suddenly able to quicken the embryo of a whale and to resurrect hind legs?'

The course and processes of evolution, and the action of the genes that control them, are still largely mysteries to us. Their study is a fascinating but rigorous discipline whose pursuit will slowly, but surely, improve our understanding of the world about us. The rest of this book will deal with living whales and how they react to their environment.

Chapter 2
The Great Whales –
the Rorquals

The first great whale that I saw at close quarters was the greatest of them all – a blue whale. The immense carcase lay on its side on a wooden deck, the flensing platform or 'plan' at Leith Harbour whaling station in South Georgia. Hulking Norwegian flensers, dressed in yellow oilskins and leather plan boots with spiked heels to keep their balance on the greasy surface of the plan, walked to and fro with hockey-stick-shaped flensing knives, giving a deft touch here and there to free the blubber so that it could be wrenched off by the steam-winches in huge strips, just as one would peel a banana.

As a zoologist, there was much to interest me there, but as a human being I was overwhelmed by the size of the animal and the shocking realisation that all those tons of flesh and bone were dead. Later I was to see many hundreds of dead whales and though inevitably one becomes hardened to

A 27.4-m (90 ft) blue whale, *Balaenoptera musculus*, on the flensing platform at Leith Harbour, South Georgia, 1953.

the sight, my fascination with these great creatures never left me, and it was their size, above all, that fascinated.

The blue whale, *Balaenoptera musculus*, is the largest creature in the world. What is more, it is the largest creature that has ever lived and would make even the largest of the dinosaurs look small in comparison. One searches for records on the subject, but, of course, it is no easy matter to measure the length of a whale, let alone weigh it. That invaluable work, *The Guinness Book of Animal Facts and Feats*, which collects together the greatest and the least, the shortest and the tallest, and so on, tells us that the largest accurately measured blue whale on record (the length being taken in a straight line from the point of the upper jaw to the notch of the flukes) was a female caught near the south Shetland Islands in March 1926. This huge creature had a length of 33.27 m (109 ft 4 in). About the same time and near the same place, a male measuring 32.64 m (107 ft 1 in) was taken.

Blue whales in the Northern Hemisphere tend to be a little smaller than those from the south. Nevertheless a 30-m (98 ft) specimen, probably a female, swam into the Panama Canal from the Caribbean on 23 January 1922. For some unexplained reason, but all too characteristically, it was killed by machine-gun fire. The carcase was towed out to sea and abandoned, but drifted ashore again at Nombre de Dios Bay. It was towed out again and the US Army Air Force used it for bombing practice. Parts of this whale (the second and third cervical vertebrae) were rescued from the shore by an English explorer and big-game fisherman, Mitchell Hedges, who presented them to the British Museum (Natural History).

In terms of weight it is more difficult to give reliable accounts of whales. It is unlikely that whales will be available for weighing in those places, like docks, which have the necessary machinery for measuring weights in excess of 100 tonnes. In consequence, whales have had to be weighed piecemeal, with all the opportunity for loss of blood etc., and the cumulative errors that this involves. The first blue whale to be weighed in this way was a 23.5-m (77 ft) specimen caught off Newfoundland in 1903. This turned the scales at 63 tons. This, clearly, was a rather small specimen. A 29.4-m (96 ft 9 in) very fat female blue whale was calculated to weigh 163.7 tons, judging from the number of boilers that were filled with meat, blubber and bones when the creature was dismembered at Prins Olaf whaling station in South Georgia. According to Sigurd Risting, a Norwegian chronicler of whaling, another very fat female blue whale, caught at Walvis Bay in Namibia, yielded 305 barrels of oil, which, at six barrels to the ton, is equal to nearly 51 tons! Unfortunately, no attempt was made to weigh this whale, but Risting calculated that to have produced this much oil it must have weighed at least 200 tons! Such a weight seems almost incredible, but Risting was a reliable recorder, and there is no reason to suppose that the oil-yield figures were exaggerated.

One of the best documented blue-whale weighings was carried out under the direction of Lieutenant-Colonel Winston C. Waldon of the US Army in 1948. If one should enquire why an American colonel should have got mixed up in the business of weighing whales, the very reasonable answer is that he was General MacArthur's personal representative in charge of the

First Fleet of the 1947–48 Japanese Antarctic Whaling Expedition, which, like almost everything else Japanese at that time, was being run by MacArthur as a sort of private empire. The whale weighed 134.2 tons, including an estimated 8.7 tons for the blood and stomach contents. It was, according to Waldon, a lean specimen: 'Larger and heavier whales were caught by the expedition, but they were not officially weighed and recorded'. It took 80 men three and three-quarter hours to flense, dismember and weigh this whale, while 25 others toiled below, processing the blubber, meat and bones on the factory deck.

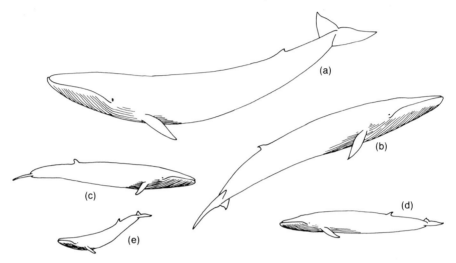

Fig. 2.1   Rorquals: (a) blue whale, *Balaenoptera musculus*; (b) fin whale, *B. physalus*; (c) sei whale, *B. borealis*; (d) Bryde's Whale, *B. edeni*; (e) minke whale, *B. acutorostrata*.

Perhaps I have dwelt too long on this question of size, but when scientists come up against something odd like this, they tend to ask why? and how? In biology it is usually more difficult to answer the whys? than the hows?, so I shall take the second part of this question first.

The largest living land animal is the bull African elephant, which, when full grown, averages around 5.8 tonnes. A record-sized bull was estimated to weigh 10.7 tonnes (24,000 lb). This is the famous Fenykoevi elephant, whose stuffed hide is seen striding across the hall in the Smithsonian Institution in Washington, DC. This was certainly a big elephant, but even the Fenykoevi specimen is puny compared with a blue whale (Fig. 2.1).

There are limits to the size that an animal living on dry land can reach. The weight of a terrestrial mammal has to be supported by its limbs, and this support, like the surface-to-volume ratio discussed in the previous chapter, is subject to a geometrical relationship. The weight of the body to be supported increases as the cube of the linear dimension, but the strength of the limbs, as compression struts, increases only as their cross-sectional area, which is related to the square of the linear dimension. Consequently, as the size of the animal increases, the weight of the body tends to overtake the strength of the limbs to support it. If we look at a series of similar-shaped animals of different size, antelopes, for example, we find that the larger members have proportionately thicker legs. If we examine the very largest land mammals, the elephants, rhinoceroses and hippopotamuses (the latter, though largely aquatic, feed on land and are actually quite nimble), we find they have a characteristic type of pillar-like leg, known to anatomists as a 'graviportal limb'.

There is clearly a limit to the extent to which graviportal limbs can compensate for increasing body weight. The great terrestrial reptiles of the Cretaceous period must have approached, if not reached, this limit. It is not merely the capacity of the limbs to support the body at rest that is important, there is also movement to be considered. A very large animal develops more kinetic energy in movement and some of this energy is transmitted up the limbs at the end of each stride, exerting stresses on the skeleton. The bones and joints of an elephant's leg have to have great strength to resist the shock of its massive footfalls. An elephant can gallop, but, should it stumble, it runs a much greater chance of injuring itself than would a smaller animal.

Whales live in a medium near their own specific gravity and hence are supported on all sides – the limitations of weight are meaningless to them unless they should be so unfortunate as to strand on shore, an event which usually proves fatal. Locomotion, for a whale, does not involve contact with an unyielding substrate, hence this limitation is also absent.

But the absence of limits does not imply that the descendants of the animal will automatically evolve to become larger. Some selective advantage of greater size must have operated so that the blue whale and its relatives, the rorquals and right whales, became 'great whales'.

Large size offers a biological advantage in a number of ways. It confers protection from many predators, which may be limited by the size of prey that they can tackle. It gives a direct sexual advantage to those species which fight for their mates (in which case we expect to find the males being significantly larger than the females). It reduces the surface-to-volume ratio, which, as we have seen in the previous chapter, is of importance to cetaceans in maintaining their heat balance. But these are general advantages, some of which at least would apply to all whales, yet we do not find that all whales are as large as the blue whale and its relatives.

Perhaps the answer lies in the way these whales feed. All whales are carnivorous, but the two main groups of whales, the toothed whales and the baleen whales, feed in very different ways. Toothed whales are, for the most part, rather conventional feeders. They eat squid or fish, usually between one hundred and one thousand times smaller than themselves, and, although we know little of their actual feeding behaviour, it is reasonable to assume that they catch their prey by pursuit, grabbing the fish or squid one by one. (An important exception to this may be the sperm whale, and I shall have more to say about this later.) This method of feeding demands a certain degree of agility, and this sets limits on the ultimate size reached by these animals.

But the baleen whales, or mysticetes, feed very differently. Mysticetes browse or graze through the great floating shoals of plankton, filtering out, by the tens or hundreds of thousands at each mouthful, prey items that are between a million and a hundred million times smaller than their predators. Filter-feeding is one of the basic feeding methods in the animal kingdom, and there are some groups, the sponges or rotifers, for example, that feed in no other way. But so unfamiliar is this method of feeding in the mammalian series that it is sometimes difficult to think of the mysticetes as carnivores. Their prey, though living animals, is passive; feeding involves no pursuit, merely the location and exploitation of a concentration of suitable plankton. Some whales have their own ways of increasing the concentration of a shoal of shrimp or young fish, but this does not amount to the hunting techniques one usually associates with carnivorous mammals.

With no requirement to pursue their prey, baleen whales have been able to sacrifice agility to the energetic advantages of great size, culminating in the blue whale.

Balaenopteridae

But it is time to look more closely at the several whales that make up the subject of this chapter. The rorquals are those baleen whales that are characterised by the presence of conspicuous grooves, or pleats, on their throats and bellies. It is from these grooves that they get their group name, for rorqual is derived from the Old Norse *rørhval* or 'grooved whale'.

Together the rorquals form the family Balaenopteridae. There are two genera: the first, *Balaenoptera*, consists of five species (perhaps six) of rather slender whales (Fig. 2.1). The largest of these, of course, is the blue whale, *B. musculus*. The Latin name seems singularly inappropriate, since *musculus* means a little mouse; it also means a muscle, but it is not clear why Linneus, who named this species in 1758, should have referred to its muscularity. As we have seen, the blue whale is very large, exceeding 30 m (98 ft 5 in) in the Southern Hemisphere, although in the north they tend to be smaller; females not usually exceeding 28 m (92 ft) or males 24 m (78 ft). In colour they are, appropriately, a dark slaty blue, mottled with paler greyish patches. The underside and the flipper tips tend to be paler. After a prolonged sojourn in cold water the blue whale can acquire a film of minute algae, or diatoms, on its skin. The pigment in these tiny plants may at times be a bright enough yellow (though more usually orange) to justify the name 'sulphur-bottom' that was at one time used for this species. There is a variant of the blue whale, first described in 1961, and generally recognised as the subspecies *B. musculus brevicauda*. This is found in parts of the Southern Ocean, particularly around the archipelago of Kerguelen in the southern Indian Ocean. It is known as the pygmy blue whale, although, despite the name, it is still a large whale, with females reaching 24.1 m (79 ft) and males 21.6 m (71 ft).

Blue whales, like other rorquals, have a cosmopolitan distribution and were formerly very common in the cold and cooler waters of both

A fin whale, *Balaenoptera physalus*, raises its flukes out of the water before submerging for a dive.

hemispheres. There are separate northern and southern stocks, for their pattern of breedng, as we shall see, keeps them apart.

The next rorqual in size order is the fin whale, *B. physalus*. This is named after its prominent dorsal fin. Although smaller than the blue whale, females reaching 24 m (78 ft) and males a little less, its fin is larger. The colour pattern of the fin whale is unique and curious. It is dark grey to brownish-black on the back and sides, with no mottling as in the blue whale (though there may be large numbers of pale oval scars scattered over the surface of the body). Beneath, including the undersides of the flukes and flippers, it is white. On the head, the coloration is markedly asymmetrical, the dark hue reaching further down on the left side than on the right. The lower lip on the right side, the mouth cavity and the anterior one-third to one-fifth of the rows of baleen plates are yellowish-white; the remainder of the baleen is striped with alternate bands of yellowish-white and bluish-grey. Occasionally, and particularly in northern specimens, it is possible to discern a pale grey chevron with the arms pointing towards the tail, and leading backwards from the eye, a dark line sweeping up towards the back and a paler one arching over to the insertion of the flipper. These latter details, however, are usually very difficult to distinguish and contribute little to the general appearance of the whale.

The fin whale is difficult for the average observer to separate from the blue whale when seen at sea (indeed, it is difficult to distinguish most rorquals), but the former has a more ridged back and, when seen from above, the outline of the head is V-shaped in the fin whale, as opposed to U-shaped in the blue.

Smaller than the fin whale is the sei whale, *B. borealis*. This gets its name from Norway, for it tended to appear off the Norwegian coast at the same time as the coal-fish, or *sei*(pronounced something between 'sigh' and 'say' in English). Sei whales can reach up to 20 m (66 ft) in length, but females average about 16 m (53 ft) and males about 15 m (49 ft). They are dark grey on the back and sides and the hind part of the underside. There is an area of greyish-white in the region of the ventral grooves. These are fewer than in the blue or fin whales, numbering about 38–56, and end well in front of the navel. The baleen plates, like those of the blue whale, are black, but differ in having a much finer, greyish fringe. On the top of the snout there is a single bony ridge. Sei whales are cool-water whales, avoiding the very coldest oceans, which the blues and fins prefer.

A minke whale, *Balaenoptera acutorostrata*, surfaces to blow.

A minke whale, *Balaenoptera acutorostrata*, rolls through the water near the ice-edge off the Antarctic Peninsula.

A minke whale, *Balaenoptera acutorostrata*, trapped in an ever-narrowing hole in the ice in the Antarctic. Freezing seas present a danger to aquatic mammals like this whale and the crabeater seals in the background, which must have access to the air to breathe.

35

A minke whale, *Balaenoptera acutorostrata*, surfaces and exhales. The spray from the blow comprises drops of water and smaller droplets condensed on minute nuclei of the oily foam that is found in the air spaces of the upper respiratory tract of whales.

Very like the sei whale, and only recognised as a separate species in 1913, is Bryde's whale, *B. edeni*. This was named after Johan Bryde (pronounced something like 'breuder'), a Norwegian who built the first whaling station at Durban in South Africa. Despite claims in a recent field-guide, it is very difficult even for an expert to distinguish Bryde's and sei whales at sea. With the whale on shore it is simple enough, for Bryde's whale is slightly smaller (12.5–14 m or 41–46 ft) and has three prominent bony ridges on the top of the snout. Even more confusingly, there may be two forms of Bryde's whale, a slightly larger offshore form (sometimes known as *B. brydei*, which may form a sixth member of the genus) and the better known more coastal form. Bryde's whale frequents warmer waters than the other rorquals described here, and there may be some populations that spend the whole of their lives in tropical waters, instead of undergoing the feeding migrations to cold waters, characteristic of the majority of rorquals.

The final, and smallest, *Balaenoptera* is *B. acutorostrata*, the piked rorqual, or minke whale. Minke (pronounced 'minkeh') is another Norwegian name (the influence of Norwegian whalers pervades the whole of the study of whales). It is said (although I am not sure it is on good evidence) that the name derives from a Norwegian whale-gunner named Meincke (which would be a very unusual name for a Norwegian) who mistook these small whales for blue whales! Be that as it may, minke whale has come to be the accepted name for this species, except in Norway, where they are referred to as '*vaagehval*' or 'bay-whales'.

Minkes are by far the smallest of the rorquals, reaching only 9.4 m (31 ft) for females and 8.2 m (27 ft) for males. They are less slender than the other members of the genus and have a sharply pointed triangular snout, from which the name 'piked rorqual' derives. They are generally black above and pure white beneath, from the chin to the tail flukes. Sometimes there is a conspicuous white band across the middle third of the upper surface of the flippers. This is a good field identification characteristic, but, annoyingly, it is often absent in Southern Hemisphere minkes. Often there are areas of light grey on the flanks, one just above and behind the flippers and the other in front of and below the dorsal fin. Occasionally there may be pale grey bracket marks, looking like gill-slits, on the side above the flipper, but these are often difficult to distinguish.

Minke whales are the most abundant of the rorquals, with a population currently estimated at about 505,000. They have a very wide distribution, showing a characteristic migration from cold polar feeding grounds to tropical or sub-tropical breeding grounds.

The remaining rorqual genus, *Megaptera*, contains only a single species, the humpback whale, *M. novaeangliae*. This is such a fascinating animal, and has attracted so much popular attention, that I shall consider it separately in Chapter 4 with another mysticete, the grey whale, which lies somewhere between the rorquals and the right whales.

Feeding

All the rorquals feed in a similar manner; the differences between them seem minor, but in fact they are of quite profound ecological significance and may be the main mechanism that allowed the whales we see today to evolve as separate species.

In all these whales, the head is very large in proportion the body. This is to accommodate an enormous mouth. Within e filter mechanism is composed of a series of structures uniqu ween whales. Hanging from the upper jaw, on either side, is a series of elongated triangular horny plates, their inner edges frayed into a fringe of coarse or fine filaments. The fringe of one plate overlaps that of the one next behind it, to form a fibrous mat over the gaps, about 6 mm ($\frac{1}{4}$ in) wide, between one plate and the next. The plates themselves do not lie in one plane, but are curved slightly backwards and are also twisted on their long axes.

The number of plates in a side of baleen varies. In a blue whale there may be from 270–395 on each side, in a fin, 262–473. The plates are longest near the back of the jaw, and taper away to tiny ones at the front. The longest plates in a blue whale are between 95 and 120 cm (37–47 in) long; in a fin less than 92 cm (36 in). The little minke whale's baleen is only 20.5 cm (8 in) in length, but the 260–300 plates are very finely fringed.

A block of baleen plates from a minke whale. The overlapping fibres of the frayed inner edges form the filter bed that separates plankton from sea water.

The inner edges of the baleen plates do not meet in the centre of the roof of the mouth, but only a narrow strip of hard palate is left exposed (Fig. 2.2). Beneath, the cavity of the mouth is floored by the tongue. This is an extraordinary organ. Neil Mackintosh, a British marine biologist, who did some of the pioneering work on rorquals in the Southern Hemisphere, described it aptly as having the consistency of a collapsed balloon partly filled with jelly. Nevertheless, we shall see that this strangely amorphous tongue works in a remarkably precise way. The ventral surface of the jaw, as we have seen, and the belly behind it, are furrowed by the grooves that characterise rorquals. If one looks carefully at these grooves it is as though the integument has been accordion-pleated. Sliced across, the skin in this region shows, beside the blubber layer, a series of layers of muscle, giving the appearance of streaky bacon.

These are the essentials of the feeding apparatus of the rorquals. To use them, the whale has first to locate a sufficiently dense shoal of plankton. We do not know how this is done. The whale's instinctive migratory pattern takes it at the appropriate time of the year into waters where plankton may be abundant. But plankton is very patchy and it is not enough for the whale to be in the right area, it has to be in the precise spot. It may be that a sense of smell helps in this. Toothed whales have no olfactory sense, for there is no olfactory organ, olfactory nerve or olfactory lobe in the brain. This was assumed to be the case with baleen whales also, but it now seems probable that their olfactory nerves and bulbs, which are certainly present in foetal life, persist in a functional state into adulthood. Professor A. J. E. Cave has

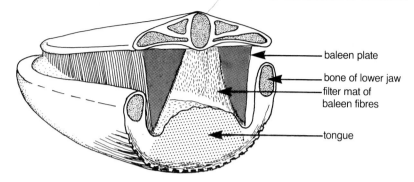

Fig. 2.2 A diagrammatic section through the head of a rorqual. The baleen plates, hanging from the upper jaw, form a sieve at the sides of the mouth, floored by the tongue.

described the functional sensory epithelium in the olfactory chamber of baleen whales and has concluded that they rely on smell to locate their food.

Smell is a sense closely allied to taste. Perhaps whales also taste the water to detect the flavour of their prey. Again, we know very little of this sense (indeed, this is true of the sense of taste in most mammals), but it is likely that the tongue does have some taste organs on it. A puzzling feature of the whale's mouth is the presence at the anterior end of the palate of a pair of shallow curved grooves. These constitute Jacobson's organ, or the vomero-nasal organ. This is a primitive sense organ and we are most familiar with its use in snakes. As a snake slithers over the ground, it constantly flicks out its forked tongue, which is then drawn back into the mouth. What the snake is doing is sampling the air, or the ground, for scent of possible prey, using the paired tips of its forked tongue. These are then withdrawn into the mouth and inserted into the paired Jacobson's organ, where the samples are tested. Mammalian Jacobson's organs are not so complicated, but in some species they certainly function as a means of smelling food already in the mouth. The puzzling thing about the whale's Jacobson's organ is that it seems to have no sensory epithelium associated with it, and would thus appear to be non-functional. I feel that more careful anatomical research might tell us much about the sense of smell and taste in whales. Alas, the opportunity was missed when fresh whale material was available in almost unlimited quantities on the factory ships and whaling stations. Who now could justify killing a whale to examine its Jacobson's organ?

We have seen that baleen whales, though virtually hairless, do retain a few hairs on the head. Hairs are important tactile organs in many mammals – think of the use a cat makes of its whiskers – and it is likely that the whale has retained its few hairs for just that purpose. Long whiskers would be a disadvantage in the water, and the whale's hairs, which protrude about 12.5 mm ($\frac{1}{2}$ in) are probably sufficiently long to pick up vibrations in the water and convey them to a ring of nerve endings deep in each hair follicle. It seems likely, although we shall probably never be able to prove it, that these tiny hairs, scattered over the lips and chin of the rorqual, act as vibration receptors and enable the whale to keep its mouth in the midst of the kicking plankton as it ploughs its way through the swarm.

Whatever sense, or combination of senses, is used, whales are successful in locating their prey. While moving forward slowly, opening the huge mouth is sufficient for it to be filled with water (Fig. 2.3a). The two halves of the lower jaw are loosely joined in front and their articulation with the skull allows them to rotate so as to widen the space between them. The accordion-pleating of the throat and belly blubber expands; the tongue is

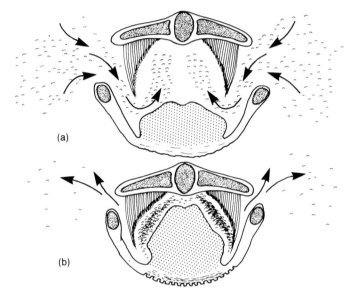

Fig. 2.3   Feeding in a rorqual. In (a) the mouth is opened and water, containing plankton, is drawn in. In (b) the tongue expands to fill the mouth cavity, expelling the water through the filter bed of baleen fibres, and retaining · the plankton.

driven backwards into a space under the chest to form a vast hollow sac, thus increasing the cavity of the mouth still further, and the floor of the mouth balloons out. At a single gulp the blue whale can take in some 70 cubic metres (2,450 cubic ft) of water, or 60 tonnes!

If the whale has judged it correctly (and we can be pretty sure it has), this water will be full of teeming plankton organisms. The jaws close by contraction of powerful masseter muscles with the floor of the mouth and the throat still distended, and then, by contraction of the muscles around the pleats, these are drawn together, reducing the volume of the mouth cavity (Fig. 2.3b). This forces water out through the filter bed of baleen fibres and between the baleen plates to the outside. Particles too large to pass through the filter are retained on the baleen fibres. The mouth cavity continues to reduce in size as the pleats continue to contract and the tongue is forced forward to resume a more conventional position in the mouth.

This leaves a semi-solid mass of plankton trapped on the filter bed in the mouth. How, precisely, this is passed into the gullet we do not know. The conventional way for mammals to swallow their food is by muscular action of the tongue. A human tongue is amazingly manoeuvrable. It consists of little more than a pad of complex muscles, with fibres running in several directions, the whole enclosed in a tough epithelium; but without thinking about it we can swallow food, locate and spit out a cherry stone, or probe a hollow tooth. A rorqual's tongue seems very ill constructed for such precise manipulation and yet there is no doubt that whales do swallow their food successfully. Like so many facets of their lives, this is one about which we know very little, and even the devising of experiments to find out more seems beyond us!

Before going on to look more closely at the various feeding patterns, and food, of the rorquals, it is worth considering their baleen more closely. Baleen consists of keratin, the tough flexible protein that makes up the superficial layer of the skin, or our nails, or the horn of a rhinoceros. But baleen seems unique in the mammalian series, not least because of its odd position, hanging from the roof of the mouth. But perhaps this is not so odd if we look at the development of the baleen in the foetal mysticete.

In a very young foetus there is no trace of baleen. There are, however, traces of teeth, and recognisable tooth-germs can be found embedded in the gum. Interestingly, the posterior tooth-germs have tri-lobed crowns, reminiscent of the lobed cheek-teeth of the archaeocete ancestor. These tooth-germs never develop. Instead, the baleen plates arise as thickenings of

Fig. 2.4   The growth of a baleen plate from a dermal process in the roof of the mouth.

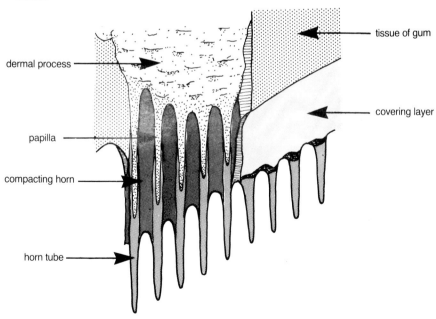

the skin on the margins of the upper jaw, to form a series of swellings marked by diagonal rows of conical processes. These processes fuse to form a series of buds or papillae, richly supplied with blood vessels. From the papillae an epidermal covering grows out, becomes cornified and forms a 'horn-tube'. At the same time, the spaces between the papillae produce a rather less dense horny substance that binds the horn tubes together (Fig. 2.4) to form a baleen plate.

It is the development of the baleen plates that gives us a clue as to their evolutionary origin. Many mammals, the sheep and dog for example, or even, to a lesser degree, ourselves, have a series of horny ridges on the palate. Baleen plates represent a greatly hypertrophied and keratinised version of these. It is difficult to visualise the evolutionary stages that could have led to the development of so highly modified and complex an apparatus. Perhaps the mysticete ancestors made their living sifting through bottom deposits for invertebrates (the grey whale does something like this today); a roughened horny palate would be an advantage for this.

Once the filter system became efficient, a new world opened up for the mysticetes. Plankton has probably always represented a majority of the biomass in the sea, and the ability of a large organism to crop this efficiently had, until the advent of the mysticetes, existed only in the whale sharks and the manta rays. It is worth noting that these, too, like the mysticetes, are the largest members of their class.

At birth the young whale has very soft and flexible baleen, but this soon hardens. The baleen continues to grow throughout life, abrasion on the inner surface wearing away the softer outer covering to release more horn tubes, which make up the fibres of the filter bed.

There are two basic methods of feeding in mysticetes – gulping, which is characteristic of most of the rorquals, and skimming, which is characteristic of the right whales, although the sei whale is also a skimmer. Gulping corresponds closely to the process previously described, though the ways in which the whale approaches its prey to take a gulp may differ. Skimming is a relatively straightforward process. The whale glides forward through the water with its mouth slightly open, causing a current of water to flow gently into the mouth and out through the filter bed, leaving the plankton trapped behind. When sufficient have accumulated, the whale swallows what it has in its mouth. Skimmers mostly feed on smaller prey than gulpers.

In this account of feeding I have consistently referred to the baleen whales eating plankton. This is not always the case. Plankton is the term used by marine biologists to refer to all those living organisms that float in the sea, drifting where the currents take them. The animals of the plankton possess, in varying degrees, the power of locomotion, but, because of their small size, this does not, for the most part, allow them to do much more than passively drift. Larger animals, that have more effective locomotion, are referred to as nekton, the swimmers. A herring is a member of the nekton; so, too, is a whale. Some of the larger members of the plankton, like the famous Antarctic krill, are capable of quite active swimming, and are sometimes referred to by purists as 'micronekton'.

Krill deserve some further description, so important are they to whales.

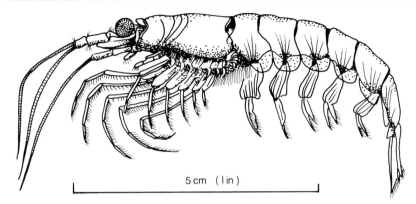

5 cm (1 in)

Fig. 2.5   Antarctic krill, *Euphausia superba*, a dominant link in the Southern Ocean food chain, and the main food, in the Antarctic, of baleen whales and many seals and penguins.

Krill is often used, in a general sense, to refer to the small crustaceans on which baleen whales feed in all the oceans. I prefer to keep the term only for those shrimp-like animals that are classified in the group Euphausiacae. Of these, the Antarctic krill, *Euphausia superba*, is by far the most important. When full grown, Antarctic krill may be up to 6 cm (2.4 in) in length and weigh over a gram (0.04 oz). Although having a general shrimp-like appearance (Fig. 2.5), krill are not closely related to the familiar shrimps and prawns that grace our tables. Krill are found in the frigid waters of the Southern Ocean in astonishing abundance, and, just as important for the whales, they occur in swarms. Swarms of krill vary in shape and size. A patch of krill may be a few tens of metres across, or the concentrations may extend over hectares. A super-swarm of krill, estimated to contain more than 2.5 million tonnes of these small shrimps, was observed near Elephant Island in February 1981. Krill feed mainly on algae, filtering the phytoplankton like miniature whales, although they are great opportunists and can eat detritus, or even each other. They form the key organism in the complex Antarctic food web, and provide the main food source not only of the baleen whales, but also of the very abundant crabeater seals, the penguins and many fish and squid. Perhaps as much as half the energy that sustains the Southern Ocean food web is channelled through krill.

There are other euphausiids besides krill. Indeed, euphausiids are one of the dominant forms of the zooplankton. In northern waters a slightly smaller, but very similar, form is *Meganyctiphanes norvegica*. Calanoid copepods are smaller than krill, but they too are eaten by whales. The sei whale and the right whales, all skimmers, eat many copepods.

In the Southern Ocean fish are more or less replaced by the huge krill swarms, and they are unimportant in the diet of whales there, but in the northern waters shoaling fish, like herrings or capelin, are a main item of diet of many mysticetes.

Food selection in whales is far from random. Different whale species, and different age-classes of the same species, feed on different-sized prey. There

43

is some correlation between the size of prey taken and the fineness of the baleen fringes, and the rorquals can be placed in a series in order of decreasing prey size and increasing fineness of baleen: blue, fin, humpback, Bryde's, minke and sei. This is not an absolute ranking, for in the Southern Ocean, where krill is so abundant, all species will feed on krill, while a young blue whale, with finer baleen, may take smaller prey than a mature fin.

David Gaskin, a whale biologist working in Canada, has given a good account of feeding in the fin whale. One August, when in a boat equipped with special devices to avoid sound and vibration that might scare whales, he was fortunate to come up with a group of fin whales off the south-western coast of Nova Scotia. Here, during the summer, swarms of copepods and euphausiids are preyed upon by schools of herring and mackerel. The swarms of plankton are aggregated over the offshore banks and ledges by the action of the strong tidal currents of the Bay of Fundy, and here, shortly after the tide turns, the fin whales begin to gather.

The euphausiids, mainly *Meganyctiphanes*, were being constantly harassed by mackerel, and formed dense layers in the upper one or two metres. The fin whales took advantage of this situation. Typically, a whale would advance on a concentration of euphausiids by making a shallow ascent, clearly visible beneath the water. As the whale came towards the surface, it would roll over on its side, usually the right side, while at the same time opening its mouth. With the mouth open at an angle of about 30° or more, the whale would turn in a tight half-circle in barely its own body length, rotating as it did so, in order to bring it back once more on an even keel with its mouth closed. It appeared to Gaskin that this manoeuvre served to drive the prey across the whale's turning circle, or at least to enable it to scoop up a mouthful as the prey streamed past its open jaws. Swimming on its side like this enabled the whale to change direction more rapidly than would be possible were it on an even keel, since the degree of lateral flexibility in the whale's backbone is small. At the peak of the feeding lunge more than half of the head and the whole of the left (usually) flipper were exposed above the surface.

A pair of American cetologists, Bill Watkins and Bill Schevill, watched fin whales feeding on fish off Massachusetts. The whales approached the fish just below the surface, keeping their mouths closed until they were just behind the fish, when the jaws were opened wide, with considerable distension of the throat, and the fish were engulfed. On another occasion, when the whales were feeding at some 15–20 m (49–66 ft) depth over a sandy bottom, they were seen to roll over on their sides as they approached the fish. However, Watkins and Schevill did not think that the whales were actively herding the fish, as they do euphausiids.

The object of all this feeding, of course, is to obtain energy and nutrients for growth and maintenance. Working out energy budgets is a familiar operation in livestock rearing, where rations have to be matched very closely to growth, and there is a lot of literature on the subject. To try to apply the same techniques to as intractable a subject as a great whale would seem to be a futile task, yet this is precisely what Christina Lockyer, a British whale scientist, has done.

She worked on the material that became available at whaling stations – statistics gathered from thousands of carcases from the great Southern Hemisphere whale fishery – and observations that she was able to make herself on a smaller number of specimens at the Hvalur whaling station in Iceland. The enormous size of the carcases presented her with major difficulties in handling, even with willing assistance from the factory staff at the station, as she made multiple measurements of blubber thickness and the weight of various organs and tissues and took samples for biochemical analysis.

The Southern Hemisphere whales provide a simpler picture, for they feed almost exclusively on Antarctic krill, and a great deal of information is available on features like body dimensions, weight and fat content of tissues, all of which are needed for calculating energy budgets. Additionally, they show clearly defined migratory movements, travelling south in the spring to feed on the shoals of krill in the Antarctic and then moving north in the autumn to return to their wintering and breeding grounds in warmer waters.

On average, blue and fin whales spend about 120 days a year on their feeding grounds in the Antarctic, during which time they put on an astonishing amount of weight. This was soon noticed by the whalers, who found that the first whales of the season, shot as they still moved south, were lean specimens, producing little oil, while those taken on their way north were much fatter. A blue whale can increase its weight by half during the course of a season's feeding; a fact that accounts for the wide discrepancy in weights for whales of about the same length.

This increase in weight is largely, but not exclusively, in the blubber, where oil droplets are deposited in the fat cells. Oil is also stored in the very porous bones, which can contain up to 56–69 per cent oil, or some 10 per cent of the total oil content of the whale. It is interesting to speculate what this vast store of fat does to the specific gravity of the whale. Because oil is lighter than water, the whale, at the end of the feeding season, must be more buoyant than at the beginning.

In general, rorquals are slightly heavier than water and tend to sink when they die, so that whalers were obliged to inject them with air to make sure that the carcases floated. But there are accounts of whales, taken at the end of the season, which floated without inflation, probably as a result of their increased fat stores. It would not pay a whale to become too buoyant, however, for then it would have to expend energy to remain below the surface to feed. Perhaps there is some compensation in the form of muscle growth. This shows a greater percentage weight increase than either blubber or bone (the bone, in fact, becomes lighter as fat replaces tissue fluid in its interstices).

Determining the amount of food taken daily by a whale is obviously not possible by direct observation. Records of the incidence of whales killed with food in their stomachs will give an indication of the frequency of feeding, however, and a measure of stomach capacity will show how much can be eaten at each meal. There are many accounts of the amount of food in whales' stomachs, but this is not an easy quantity to measure, and the

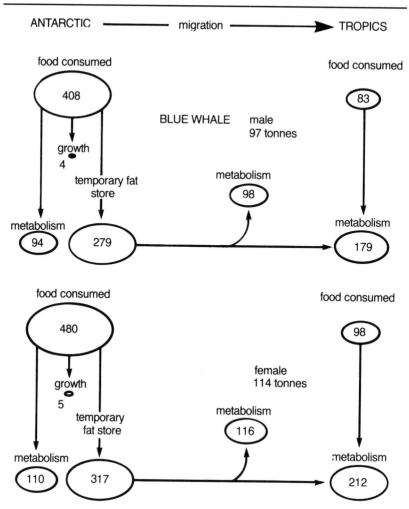

ANTARCTIC ——————— migration ——————→ TROPICS

food consumed

408

BLUE WHALE   male
97 tonnes

growth
4

temporary fat
store

metabolism
98

metabolism
94   279

food consumed

83

metabolism
179

food consumed

480

female
114 tonnes

growth
5

temporary
fat store

metabolism
116

metabolism
110   317

food consumed

98

metabolism
212

Fig. 2.6   An energy pathway for Antarctic blue whales. Nearly all the food taken in a year is consumed in the Antarctic, the whales living on their reserves, in the form of temporary fat stores, as they winter on their breeding grounds in the tropics. Energy units are in millions of kilocalories. (From data prepared by Christina Lockyer.)

process is complicated by the fact that whales, like ruminants, have multiple stomachs with three chambers. However, from what data are available, it seems that a blue whale might take a meal of about 1,000 kg (2,200 lb) of krill and a fin whale about 800 kg (1,760 lb). Given these meal sizes and the frequency of feeding, it is possible to estimate that perhaps about four meals are taken each day.

After considering these and other factors, such as food availability, digestion rate and calculated energy requirements, Christina Lockyer was able to calculate that the large rorquals would take about 30–40 g (1–1½ oz) of food per kilo (2¼ lb) body weight per day, or about 4 per cent of the total body weight. Over a feeding season of 120 days this would amount to the whale eating between 3.5 and 5 per cent of its body weight in a year.

Given this figure, Christina Lockyer was able, after carefully studying the mouth structure of these whales, to calculate how many gulps would be needed to obtain the daily ration. Asssuming a density of 2 kg krill/m³ water (and this could be substantially greater in some concentrations (a value of 35 kg/m³ has been recorded), a 28 m (92 ft) blue whale, which would weigh about 146 tonnes, would be able to take in 32.6 m³ of water at a gulp, equivalent to 65.2 kg (143¾ lb) of krill. This means that, to obtain its daily ration, the whale would have to take only 79 gulps, assuming that krill concentrations of at least this density could be found.

Interestingly, if these calculations are made for a series of whales of different lengths, it is found that, for the same species, the smaller the whale, the greater the number of mouthfuls it has to take to get its ration. A 16-m (53 ft) blue whale would need 130 gulps, while in the minke whales, a large specimen of 9 m (30 ft) would need 122 gulps, and a small 5-m (16 ft) minke would have to take no fewer than 355. Here, then, is a biological advantage associated with great size in these animals.

When the whales move north to their breeding grounds, they probably do not give up feeding entirely, but feed at much reduced rates, perhaps as little as 10 per cent of the gorging they do in the Antarctic. While in the warmer waters, where energy demands are less, they live on their stored fat. The animals that have the greatest need for energy at this period are the lactating females. These may need to put on as much as 60–65 per cent of their body weight during the feeding period to support the milk demands of their calves, and observations show that pregnant whales are the first to arrive at the feeding grounds and the last to leave.

The final product of Lockyer's work was a series of energy-flow diagrams for various species and classes of whales. Fig. 2.6 shows such a pathway for adult blue whales. It stands as a monument of what can be achieved by careful study of even the most unpromising material and difficult subjects. Christina Lockyer's conclusion, that the calculated assimilation efficiency for a mature female fin whale is 87 per cent, demonstrates for us what an efficient system the rorquals form as harvesters of plankton.

Chapter 3
Right Whales and Bowheads

The right whale and its Arctic cousin, the bowhead, represent an extreme specialisation of plankton feeding by the use of baleen. There are three members of the family Balaenidae. The familiar right whale, or black right whale, *Balaena glacialis*, is the most widely distributed. At one time it was to be found in all the temperate waters of the world, but, being slow moving and generally passive, it fell an easy prey to humans. The earliest whaling of which we have historical record was founded on right whales; they were, indeed, the *right* whales to catch. Some cetologists divide the right whale up into three subspecific forms, one each in the North Atlantic, the North Pacific and the Southern Oceans. Because the right whale does not stray far into tropical waters, or into the high Arctic, these three groups are genetically isolated, but is not certain that they are distinct. Nevertheless, the southern form is often separated as *B. glacialis australis*, or even as a full species, *B. australis*.

The right whale

The right whale (Fig. 3.1) is a great whale in the true sense, and even though not very long by rorqual standards, reaching only some 18.3 m (60 ft) in length, it is a very bulky whale and may weigh as much as 96 tonnes. The most conspicuous feature is the huge head, which can make up as much as a quarter of the whale's length. The massive arched lips reach up to a strongly curved rostrum from which hang about 230 pairs of dense black baleen plates. The body is stocky and rotund. Unlike the rorquals, there is no trace of a dorsal fin, which is how it receives its Norwegian name '*slettbak*', or smooth-back. The flippers show more of the digits through the integument than those of the rorquals.

Fig. 3.1   The right whale, *Balaena glacialis*.

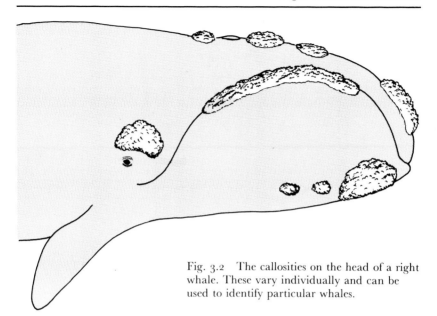

Fig. 3.2   The callosities on the head of a right whale. These vary individually and can be used to identify particular whales.

In colour, right whales are black or very dark grey, but there are often patches of white on the underside of the body. Additionally, on the head there are strange patches of hypertrophied skin, or callosities. The main one of these is known as the 'bonnet' and, appropriately enough, this sits on the top of the head in front of the blowhole (Fig. 3.2). Similar callosities may occur on the tip of the upper jaw, the end of the chin and around the eyes. The callosities are larger and better developed in males than in females. Through the callosities protrude the tips of sensory vibrissae, similar to those we noted around the snout of the rorquals. The arrangement of the callosities varies in each whale and can be used to recognise individuals. From such features Roger Payne has been able to build up a casebook of right whales which resort to the bays around Peninsula Valdes in southern Argentina to breed.

The rough and horny skin of the callosities is always infested with an interesting mixture of barnacles and strangely modified amphipod crustacea. The barnacles are purely fellow-travellers, hitching a lift on the whale, which supplies them with a substrate they can fasten to, and, by its motion through the water, a feeding current for the barnacle's arms to sift for food particles. The amphipods, which belong to the family Cyamidae, the whale-lice, (Fig. 3.3) are a little less benign. They feed mainly on the skin of the whale, but there is no evidence to show that they cause their hosts any distress. Barnacles and whale-lice are found on many whales, but principally on the slower moving kinds. Many of them are highly host-specific, being found on one species of whale only, which perhaps indicates that they are transferred from whale to whale during some intimate contact, rather in the manner of human crab-lice.

The function of the callosities, from the whale's point of view, is not

49

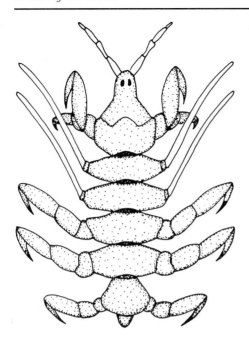

Fig. 3.3 A whale-louse, a cyamid amphipod crustacean related to the familiar sand hopper.

The right whale raises its flukes clear of the water as it dives.

A southern right whale, *Balaena glacialis*, exhales its characteristic V-shaped blow. The huge lip can be seen on the left of the whale's head.

known. Convenient as they are for harbouring the barnacles and amphipods, it seems unlikely that they have evolved for this purpose. However, Nick Arnold has reported that certain lizards develop in-pushings of the skin, which are infested with mites that feed on the epidermis, much as the cyamids do on the whale. Arnold suggested that this was perhaps an adaptive feature developed by the lizards in order to localise the mite attacks. It is just possible that the right whale might have done the same with its callosities.

Roger Payne and Eleanor Dorsey noted that sexually active male right whales showed a greater frequency of scrape-marks on their skin than females, and speculated that these might have been caused by the males using their callosities as weapons in aggressive encounters over females. Both explanations could be right, the callosities being first developed in relation to the cyamids, and then being used by the males for fighting, resulting in the greater development of the callosities in the male.

Right whales are slow, ponderous movers at sea, and are easily recognised by this feature. When they dive the smooth roll of their finless backs and the display of the flukes in the air are quite characteristic. Their blow is V-shaped, the blast diverging from each nostril, and though this is instantly recognisable to a skilled observer, it is not always very conspicuous.

Right whales are migratory, moving into higher latitudes in the summer to feed. Like sei whales, they are skimmers, but, unlike the sei, they do not have a throat pouch. There is no trace of the throat grooves that characterise the rorquals. When feeding, the right whale swims slowly forward, allowing the water current to sweep through the very fine fringes of its long baleen plates. Whereas the rorquals evolved so as to take a huge volume of plankton-containing water into a distensible throat, the right whales have developed a much larger filter bed by arching the upper jaw so that the functional length of the baleen is increased. This has necessitated

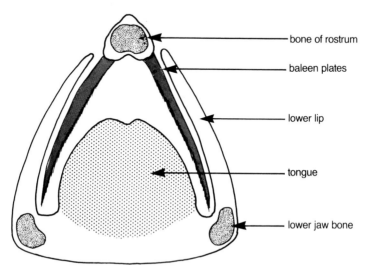

bone of rostrum

baleen plates

lower lip

tongue

lower jaw bone

Fig. 3.4   A section through the head of a right whale. The very long baleen plates and associated huge lips distinguish right whales from rorquals.

the development of the enormous lips that are so characteristic of these whales (Fig. 3.4). As the whale swims along, it is sometimes possible to hear the baleen plates rattling against each other in the water current.

Another difference from rorquals is in the tongue. In the right whale this is much smaller, with a more solid and muscular consistency. The action of the tongue in this family has not been accurately described, but it would seem to operate in a more normal mammalian manner than the tongue of a rorqual, being used to lick the plankton off the baleen. Right whales feed on smaller food organisms than rorquals, mostly taking copepods, and this is related to their 'skimming' feeding method (the sei whale, another skimmer, also feeds on smaller organisms). In the Southern Ocean, south of the Antarctic Convergence, where krill are so common, right whales feed on this, as do rorquals.

Because of their slow speed, coastal habits and general indifference to observers, right whales have been the subject of several behavioural studies, chiefly from the Bay of Fundy to Cape Cod on the eastern coast of North America, and at Peninsula Valdes in Argentina. Right whales are beginning to appear again in some numbers off the coast around the Cape of Good Hope, and this is now another area for study.

When right whales were more common, before the nineteenth-century whalers had taken their toll, they seemed to be rather social animals, with aggregations of as many as 100 whales being seen together on the feeding grounds. There is no way of knowing whether these whales were socially interacting with their fellows or whether they were simply responding to a common stimulus of a concentration of food. However, some observations I collected on sightings of right whales around South Georgia in 1959, indicated that they were seen in groups of two or more, more frequently than on their own, which leads me to suppose that they do have some kind of social relationship that is not necessarily connected with mating. While in coastal habitats, to which the whales resort primarily for breeding, right whales associate in small groups of two to nine animals. These groups do not seem to have a fixed composition, and individual, recognisable whales may move from one group to another. Groups usually break up to feed, although right whales have been seen skimming in echelon formation.

Mating occurs most often in the spring and is probably promiscuous, with no long-term bonds being established between a mating pair. On the breeding grounds a receptive female may be surrounded by a group of two to six competing males. The female may circle and dive shallowly, with the males following her, in what appears to be a courtship dance. A non-receptive female may swim away or lie on her back in the water, so that her genital region is not accessible to the males. Males may then attempt to roll her over in the water, but, as the females are larger, they are often unsuccessful in this. In any case, it is hard to believe that a really unwilling female could ever be forcibly mated against her will, as the precision of positioning required by the two huge partners in the water must demand the full co-operation of both. Without this co-operation it is very difficult to see how intromission could be achieved.

Another peculiar feature is the occurrence of 'triads' of two males and a female. One of the males appears to support the female while the other mates with her. Examples of altruistic behaviour such as this, which involves mating, are difficult to explain in terms of conventional evolution by natural selection. In some cases altruistic behaviour can be accounted for by postulating kin-relationships in the animals taking part, but nothing is known of such relationships in right whales.

Gestation lasts for nine to ten months and a single calf is born, usually in the protected waters of a shallow bay. However, this is not the case in a population of right whales inhabiting the Bay of Fundy, which mate and rear their calves in deep offshore water away from the coast. One is tempted to speculate whether the appearance of this behaviour pattern, which might not seem well adapted to the survival of the young, could be a result of human predation on whales breeding in more coastal waters. The newborn right whale calf maintains close contact with its mother, often swimming up on to her back, butting her or even covering her blowhole with its flukes. Touchingly, the mother may at times roll over and hold her baby in her flippers. One should not cynically dismiss such observations, or comments on them, as anthropomorphisms; the function of creating a bond between mother and offspring is a general mammalian characteristic and one that is especially well developed in primates and cetacea.

Much of what is known of the behaviour of right whales has been learnt in the shallow protected waters of the Golfo San José, enclosed by the Peninsula Valdes in Argentina. Here it is possible to observe right whales from vantage spots on the cliffs, as they mate or nurse their young. The ability to recognise individual whales from their callosities has allowed some observations to be made on calving intervals. These may vary from two to four years, but most females have a three-year gap between successive calves. This implies a very low reproductive rate and may account for the slow recovery of depleted right whale populations, despite many years of protection.

Right whales show a seasonal pattern of north-south migration, but it is less regular and well defined than those of some other baleen whales, such as the grey whale. They seem rarely to spend long in very cold waters, although there is a record of one off the South Orkney Islands and they are reasonably abundant (and were once very abundant) in the waters around South Georgia; both localities well south of the Antarctic Convergence. In high northern latitudes, the place of the right whale is taken by another species.

The bowhead

This is the bowhead (*Balaena mysticetus*), or Greenland right whale, which is confined to the polar seas of the Northern Hemisphere. It is similar in body shape, although it is perhaps slightly larger, lengths of over 20 m (66 ft) having been claimed. There are no callosities on the head, but there is usually a characteristic white patch under the chin. The baleen is longer, up to 4.46 m (14.6 ft) in length, and there are between 325 and 360 pairs of plates. The head is, proportionately, even larger than that of the right whale, in order to accommodate this massive feeding apparatus, making up almost a third of the length of the body (Fig. 3.5).

Bowheads are essentially high Arctic right whales. Although their

Fig. 3.5   The bowhead whale, or Greenland right whale, *Balaena mysticetus*.

natural history is not well known, on account of their rarity and inaccessible habitat, bowheads seem to live in much the same manner as right whales. They feed by skimming, taking a mixture of copepods and euphausiids, but they do occasionally feed near the bottom, probably taking amphipods from the surface of the sediment in the manner of grey whales.

Their large size may be an advantage in breaking through ice. A bowhead can break new ice up to 30 cm (12 in) thick by heaving against it with its back; presumably the larger animals can break even thicker ice. In this way bowheads can create their own breathing holes and extend their range into areas not accessible to other marine mammals. Often bowheads may be accompanied through icy waters by belugas, the smaller whales taking advantage of the breathing holes opened by the larger ones.

There are, or were, four populations of bowheads, although whether these were all genetically isolated we can no longer tell. The stock that once abounded in the north-east Atlantic, around Svalbard, is now extinct. Perhaps a few hundred remain in Davis Strait, around Baffin Island and in Hudson Bay. In the Western Hemisphere a few hundred inhabit the Sea of Okhotsk, while the main population, numbering perhaps around 3,900, are found in the western Arctic, around the Bering, Chukchi, Beaufort and east Siberian Seas. It seems likely that the initial population of bowheads, before the inception of commercial whaling, was of the order of 60,000. Of this, only around 5,000 remain and the status of even this remnant is uncertain.

The right-whaling industry

The great reduction of right whales and bowheads has been the direct result of predation by humans in the pursuit of commercial advantage. Subsistence hunting existed since prehistoric times, but probably had little effect on the stocks of large whales. Commercial hunting may have started some time before the twelfth centry with the killing of right whales by the Basques, the people inhabiting the French and Spanish coasts of the Bay of Biscay. Neighbouring villages may have combined together to cut up and render down the blubber of whales that occasionally stranded on the shore. The Bay of Biscay was probably a major breeding area for right whales, much as the Peninsula Valdes is today, and the whales doubtless came very close inshore. It would have been a simple enough task for the Basques, who seem always to have been accomplished fishermen, to urge the creatures further inshore to strand on the beach, perhaps encouraged by a thrust from a lance. From that point to killing them at sea would not have been a great departure.

By the twelfth century, Basque whaling was already economically important to the region. Many Basque towns incorporated a whaliing scene in their coat of arms; Biarritz, for example, shows a boat in the act of harpooning a whale. The boat was known as a '*chaloupe*', from which we get our term shallop, a boat used by nineteenth-century American and British South Sea whalers and sealers. The word harpoon itself is said to be derived from the Basque word '*arpoï*', of which the root is said to mean to take quickly, a singularly inept term for a weapon that, for the first eight centuries, involved men in struggles with whales that often lasted many hours. Classical scholars will notice a curious resemblance between the Basque word and the Greek verb 'to snatch'. Perhaps there is still work here for etymologists.

The Basques built stone look-out towers, of which a few ruins still remained when Sir Clemant Markham, an English geographer and a historian of Basque whaling, visited the area in 1881. When a whale was sighted, signal fires were lit and the shallops were launched. These were shallow-draught rowing boats with a harpooner, a steersman and ten oarsmen. The harpooner, standing in the bow, would hurl his weapon into the whale and then change places with the steersman, a curious and inconvenient manoeuvre that persisted right up to the end of Yankee whaling and earned the displeasure of Herman Melville, among others. When the crew got the opportunity, they hauled up on the line until they were near enough to the whale to lance it. Once the whale was killed, it was towed back to the beach and cut up.

Blubber, meat and baleen were all used, the meat being especially highly prized. The Church and the king came in for their share, according to Harrison Matthews, the Church often receiving the tongue, while the king was entitled to 'a slice of the whale along the backbone from head to tail'.

As the industry developed, it became necessary to look further afield for whales and the Basques responded by building larger vessels that could cruise offshore. By the sixteenth century the Basques were voyaging as far away as Newfoundland, mainly for the cod fishing, but also to catch right whales. A letter from an English merchant, written in 1578, reported 20 to 30 Basque whaling ships on the Grand Banks. These were caravels, large clumsy vessels of up to 700 tons. Each carried a shallop, or more than one, from which the whales were hunted. When a whale was caught, it was towed to some sheltered anchorage and stripped of its blubber, which was boiled out in a temporary try-works erected on the beach.

Right whales seem to have been reasonably abundant off Biscay, at least until the mid-seventeeth century. The total Biscay catch for the hundred years from 1517 was of the order of 700 to 1,000 whales. The industry continued in a sporadic fashion until as late as the end of the eighteenth century, although, by that time, it had been overtaken by events further north.

The next phase of whaling was presaged in 1553, when a trading mission, in search of a north-east pasasage to China, reached Archangel. On the way they sighted Spitzbergen and noticed the abundance of whales. These were Greenland right whales, or bowheads (although they were never called by that name in Europe). The first whales to be killed at Spitzbergen were taken in 1610 by Thomas Edge, an Englishman employed by the Muscovy Company. His first season was scarcely a success, as one of his two ships was destroyed by ice in a supposedly safe anchorage, while the other, coming to the rescue, stowed her cargo holds so badly that she capsized. However, part of the cargo and both crews were rescued by a third ship and taken safely back to England.

In 1612 the Dutch entered the Spitzbergen whaling region and there

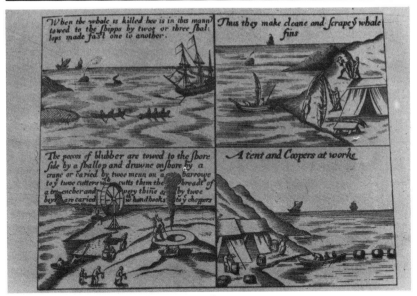

Whaling at Spitzbergen in the seventeenth century.

began a long history of conflict between the English, who claimed a monopoly they could not enforce, and the Dutch, who defended their right to whale. A sort of agreement was eventually reached between the main contenders, who, by 1618, had been joined by various other European nations, but the Dutch finally emerged as the most successful.

At this time, whaling from Spitzbergen was solely dependent on shore works. Whales were hunted in the large bays of the archipelago (Svalbard, as it should be known – Spitzbergen is really the name of the largest island) from shallops, as in Basque whaling. The main Dutch settlement, Smeerenberg ('Blubbertown'), was, at its height, a community said to have numbered some 18,000 people, with inns, bakeries, shops and chapels, as well as the buildings necessary for the whaling: try-works, cooperages, etc. However, these were 'summer-only' towns; very few people wintered in Spitzbergen and those that did (if they survived) found it an extreme hardship.

Scoresby, a nineteenth-century English whaler, recounts how some condemned criminals, for whom their employers had obtained the promise of a reprieve if they were willing to spend a complete year at Spitzbergen, on viewing the land that they were to inhabit, were so terrified by its appearance and bleakness that they begged to be sent back to the fate that awaited them. (They were sent back, but, happily, were not executed.)

Around the middle of the seventeenth century, whaling in the bays began to fall off as the stocks of whales were reduced. The Dutch, who by this time had become masters of the trade, responded by moving their hunting grounds to the pack-ice of the north, killing the whales in the open sea, and flensing them in the water alongside their ships in whatever shelter was afforded by the ice. This separated the ships from the try-works ashore, so they stowed the blubber in casks, trying it out on returning from the voyage. Whaling in the ice could be carried out over a longer season, as the whalers were not dependent on the presence of inshore breeding animals. Meanwhile, the English, who had remained bay-whalers, found their trade languishing.

Whale oil, however, was still an important item of commerce. A main market was the soap industry. Traditionally this had used tallow, but whale oil was cheaper, although it did not make such good soap. The great growth of the woollen textile industry in the eighteenth century provided another market for whale oil, which was used to scour the wool prior to spinning, particularly in the production of coarse cloths, such as the military serges that were in such great demand in Europe, at least until Napoleon's defeat at Waterloo in 1815. Another important use at this time was for street and factory lighting. Five thousand street lamps, burning whale oil, were installed in London in the 1740s. The other important product from the Greenland whale was its long and very flexible baleen. This was in great demand for stiffening ladies' garments, particularly the hooped skirts fashionable until the 1820s. Baleen, or whalebone, as it was always known, was not used only in the garment trade. Its remarkable flexibility and great strength made it suitable for carriage whips, umbrella stays, fishing rods and so forth, but perhaps its most enduring use, which continued well into the present century, was in the production of bristles for brushes.

Continued whaling around Spitzbergen led predictably to the decline of the stocks there and the whalers had then to look elsewhere. By 1820 the centre of the industry had moved to the Davis Straits, the frigid waters between Greenland and Baffin Island. Here another stock of Greenland whales awaited exploitation. The Davis Strait whale fishery was largely in the hands of the English, and later Scottish, whalers, the participation of the Dutch having declined. William Scoresby, a whale captain from the Yorkshire port of Whitby, was the great chronicler of this phase of whaling. Scoresby, in his *Account of the Arctic Regions* (1820), has left us a vivid picture of the life of whalers, and the death of whales, at this time.

Their ships were strongly built, with treble or double planking and of about 330–340 tons. Massive props in the holds were set up to resist ice pressure, but many of them still succumbed. They carried 40 or 50 crewmen and had six or seven light whale boats. Initially, the whales were killed with hand harpoons and lances, but from about 1790 harpoon guns were introduced, which could project the harpoon some 37 m (40 yd). The harpoon was connected to a hempen whale line some 1,300 m (1,417 yd) long, coiled in a tub in the boat.

When a whale was sighted from the mast head barrel, or crow's nest, the boats were lowered. They endeavoured to come up behind the whale (Scoresby believed the whale to be 'dull of hearing, but quick of sight'!) and were careful to row smoothly. The successful harpooner was one who could anticipate the movements of the whale and position his boat close to it as it surfaced. The harpoon was planted in any part of the body that could be struck, but forward of the flippers was preferred.

Once a whale was struck, it might lie quietly in the water, allowing the boat to haul up and lance it to death, but more usually it would set off at a great speed, towing the boat behind it. This was a desperately dangerous business in a sea of pack ice, as the boat would be destroyed if the whale swam under a floe. The strength of a whale was almost unbelievable. Nothing was thought of a whale taking out miles of line and towing four or five boats. Scoresby records how a medium-sized whale, in 1812, took out 9.6 km (6 miles) of line from a total of eight boats, one of which sank and was towed by the whale under the water. On another occasion a whale survived 40 hours after being struck, and for the last two hours of its life it was towing the whale ship directly to windward at a speed of 1–2 knots!

Once dead, the whale would be secured alongside the whale ship and the process of flensing, or 'flenching' as it was more often known at that time, begun. Men in spiked boots descended on to the back of the whale, while others in boats alongside would help them. Heavy cutting tackle, suspended from the mast head, heaved off a spiral strip of blubber as it was freed by the cutting-spades of the men on the whale. When all the blubber had been brought on board in this way, the head was cut off and brought aboard to be stripped of its baleen. The blubber was cleaned of any meat that might be adhering to it, and packed in casks which were stowed in the hold until

SOUTH SEA WHALE FISHERY

The South Sea Whale Fishery. This nineteenth-century print is a good representation of a trade that reduced the population of right whales in the Southern Hemisphere to vanishingly small numbers.

the ship returned to its home port. It can be imagined that a Greenland whaler was an oily, smelly vessel.

Right whaling was not confined to the north Atlantic region. In 1775 a New Englander, Samuel Enderby, arrived in London, a refugee from the American War of Independence. Enderby sent whale ships to search the southern seas, the bays of South America and island groups like the Falklands. These South Sea whalers combined right whaling with the pursuit of sperm whales and a good deal of fur seal and elephant seal hunting. Enderby's ships were joined by American whalers, operating out of New England and chiefly hunting sperm whales, although available right whales were rarely ignored.

By 1800 the South Sea whalers had penetrated to all the breeding localities of the right whales, particularly around South Africa and Tasmania where right whales were especially plentiful. Fifty years later the stocks of whales were greatly reduced and, as in the Northern Hemisphere, the trade declined. However, on account of the vastly greater areas of sea over which to hunt, whaling still continued, although on a greatly reduced scale. The last Hobart whaler was fitted out in 1894 and made several voyages with varying degrees of success, until she was withdrawn in 1900.

In the Arctic the British fishery ceased at all ports except a few Scottish ones. The jute industry, centred at Dundee, was a great consumer of whale oil, which was used in the weaving process, and this kept the Dundee fleet active. Steam was introduced in 1857, but this did not make up for the scarcity of whales. In 1902, for example, the whaler *Active* went to Hudson Strait and returned with one small Greenland Whale, 11 white whales, 54 walrus, 205 seals and 77 polar bears.

For a time fashion came to the aid of the industry when whalebone corsets came into great demand. Whalebone reached a peak price of £3,000 a ton, so that the bone from one or two whales could pay the expenses of a voyage and yield a reasonable profit. In 1905 the value of whalebone landed at Dundee was eight times that of the oil. By this time, however, the Arctic whale fishery in the Eastern Hemisphere was nearly extinct; the last two whalers to sail from a British port, the *Morning* and *Balaena* sailed from Dundee in 1913, but returned home empty.

This, however, was far from the whole story of Arctic whaling. Yankee sperm whalers had been cruising the Pacific since 1791, and in 1848 Captain Thomas Roys took his ship, the *Superior*, into the Bering Sea in the expectation of finding bowhead whales there. Roys had heard from a Russian naval officer, whom he had met in a hospital in Petropavlovsk, of great whales in these little-known waters. These expectations were amply rewarded, as Roys discovered a virtually unexploited and major stock of bowheads. The news of Roys' success spread quickly and, by 1852, more than 200 whalers were operating in the Bering Sea. The American whale fishery was exceptionally well documented, and surviving log books have been very well studied, so we know that by the end of the bowhead-whaling era in this region, in 1915, 16,600 whales had been taken and a further 2,050 had been killed but lost.

Bowheads winter in the Bering Sea, particularly around St Lawrence and

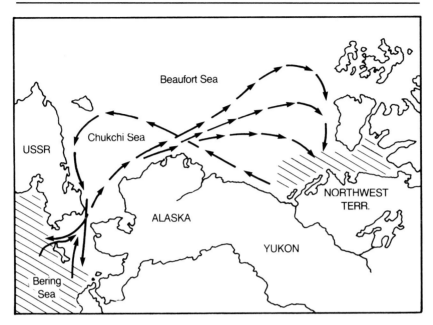

Fig. 3.6 Migration paths of Pacific bowheads.

St Matthew Islands (Fig. 3.6), which seem to be where the young are produced after a gestation period of 10 to 11 months. In the spring the animals migrate north-eastwards through the Bering Straits into the Chukchi and Beaufort Seas. The fractures in the sea ice – the leads – lie close to the shore, so that most of the population pass close to Point Barrow on their way north. As the ice retreats, the whales progress further east, reaching as far as the mouth of the Mackenzie River and Amundsen Gulf during the second half of July. The return migration in late summer or early autumn tends to be further offshore than the spring passage.

The American whalers were not the first to hunt bowheads in this region. For more than 2,000 years the indigenous Alaskan Eskimos had hunted these great creatures from sealskin-covered boats called '*umiaqs*', striking them with bone or ivory harpoons to which were attached, by sealskin lines, a number of floats made from inflated sealskins. These floats not only impeded the whale's progress through the water, they also indicated the direction of its movement, enabling the hunters to keep up with the whale and to seize the best opportunities to lance it.

All Eskimo hunting was deeply imbued with taboos and religious ceremonies, but none more so than whaling. For example, the preparations at the beginning of the whaling season involved not only cleaning and repairing the boats and whaling tackle, but also the making of complete new sets of clothes for the hunters, as for religious reasons, the whaling crew could not approach whales in clothes previously used for hunting. Prior to the hunt, the whaling crew would enter a period of sexual abstinence, intensive meditation and spiritual preparation. The equipment – harpoons, lances, lines and floats, and particularly the boat – had all to be made ready not only physically, but also spiritually. Amulets had to be prepared and placed in the *umiaq*, which was hallowed by libations of water. When, after a successful hunt, a whale had been secured and landed on the ice-edge, an elaborate ritual of greeting and benediction was performed to assuage the spirit of the whale. The wife of the whaling captain of the *umiaq* which had taken the whale would pour water from a specially constructed wooden container on to the snout and blowhole of the whale, urging it to drink and come back to the village next spring. Then the wives of the rest of the whaling crew would thank the whale for allowing itself to be taken, and welcome it to the village. Finally, the whale was cut up and divided among the whole community, according to strict rules.

The origin of such customs is obscure, but they were widespread among subsistence hunters. Pursuing a whale must have seemed an awesome task to primitive Stone Age peoples. The acquisition of spiritual power, as well as material skills, was clearly regarded as vital, and this could only be done by the observance of complicated ceremonies. Such rituals may, I believe, have had an important effect in the development of the relationship between primitive communities and the living resources on which they were dependent. The requirement of elaborate ceremonial and ritual taboos before, during and after a hunt would have restricted the hunting pressure, allowing a better opportunity for the hunters and their quarry to stay in balance and ensure that the stock was not overfished.

Such a balance probably existed between the Alaskan natives and bowhead whales prior to the arrival of the American whalers in the nineteenth century, but it was soon upset. Eskimos and commercial whalemen came into significant contact some time after 1870. The Eskimos wanted to obtain access to the technology that the Americans had brought with them, in particular the darting gun and bomb lance. The darting gun was a device that could be fastened to the harpoon shaft to fire a small explosive projectile into the whale; a bomb lance fired a larger bomb into a whale after it had been fastened-to with a harpoon. Both devices helped to kill the whale more speedily than the stone-tipped harpoons and lances used by the Eskimos at that time.

By the late 1880s the Americans had begun to set up whaling camps along the ice-edge and to employ Eskimo hunters to try to halt the decline in catches. This was scarcely possible, as the stock of whales had been so greatly reduced, but, by increasing the hunting pressure (some stations sent out as many as 20 crews in the spring), it was possible, initially, to maintain the catches. By this time the only commercial product was baleen, which suited the Eskimo hired hands very well, for they had little use for this but could retain all of the rest of the whale.

By 1908 the industry had collapsed, largely through the shortage of whales, but also as a result of the replacement of baleen by spring steel in the corset industry. The Eskimos reverted to their early subsistence hunting, but this was based on a very depleted stock of whales. In some communities whaling died out altogether, although in others the whaling crews struggled on at a catch level of between 10 and 15 whales a year.

In the 1970s the level of whaling expanded dramatically. This was the result of the resumption of whaling by three communities where it had lapsed, and also by more crews participating in those settlements which had a continuous history of whaling. Almost certainly a major reason for this was the improved economic position of the Eskimos (a result of minerals exploration on the Alaskan North Slope), which enabled more individuals to finance whaling crews. An associated factor may have been a decline in the western Arctic caribou herd, which forced the hunters to look for other sources of meat. Whale meat and caribou were, in that order, the preferred food of the native peoples. Between 1973 and 1977 totals of 47, 51, 43, 91 and 111 bowheads were struck or killed, with the proportion of those struck but lost increasing during this period. As the presumption is that a whale which has been hit with one or more explosive bombs is likely to die of its internal injuries, this represented a very considerable toll of the surviving stock.

The traditional hunting rights of the Alaskan Eskimos had been respected by the American administration, but this take, of what few people could deny was a very rare species of whale, attracted worldwide attention. The International Whaling Commission, the body set up at the end of World War 2 to control the whaling industry, had, from its inception, provided total protection for all right whales, including bowheads. The United States was a member of the IWC, and one with a high profile on the topic of whale conservation. It came as no surprise to most people when, in 1976, the IWC passed a resolution recommending '... that contracting governments as early as possible take all feasible steps to limit expansion of the fishery and to reduce the loss rate of struck whales'. The following year, 1977, the Scientific Committee of the IWC recommended that the Commission should rescind an exemption that had permitted aboriginals to take bowheads. The Committee further called for a zero quota for the bowhead catch in Alaska. The Commission acted positively on this recommendation and announced a total ban on bowhead whaling.

This decision shocked and angered the Alaskan Eskimos. They had believed (indeed, still believe) that it was their prerogative alone to decide how to harvest the bowheads that migrated past their shores. They pointed to the cultural importance of the bowhead hunt to their society, and many stated that they would continue the hunt, irrespective of the IWC's imposed ban.

This put the US Government in a very difficult position. They were reluctant to appear indifferent to the fate of the undoubtedly greatly reduced bowhead population, particularly as they had taken a strong position within the IWC in affording protection for other endangered whale stocks. Equally, they could not afford to ignore the claims of the Eskimos. Earlier treatment of aboriginal peoples on the North American continent still caused much resentment and feelings of guilt, and the USA, generally, had been a strong champion of the rights of minorities.

The rules of the IWC allowed participating governments to submit, within 90 days, a formal objection to any ruling of the Commission and then not be bound by it. If the USA entered such an objection, the ruling would not apply to it and the Eskimo whaling could continue. However, such a course was not seen as politically tenable, and the State Department withheld its hand.

Meanwhile, the Eskimo whaling communities had set up their own Alaskan Eskimo Whaling Commission (AEWC), formed from the 70 whaling captains from the various whaling settlements. The AEWC announced that it did not recognise the jurisdiction of the IWC over their hunt and would not accept the zero quota. US government officials managed to open the matter again with the IWC at a special meeting called in Japan in December 1977. The USA came forward with a management plan that would put a limit on the numbers struck, as well as landed, coupled with a plan to carry out intensive research on the bowhead stocks. This proposal was rejected on biological grounds by the Scientific Committee. With the little that was then known about the population level of the bowheads, the only justifiable course had to be total protection. Any other course would erode protection measures for much less endangered species, such as blue whales, humpbacks or grey whales. However, the Scientific Committee realistically recognised that the Commission might not wish to examine the matter on solely scientific grounds; subsistence and cultural needs, on which the AEWC had been highly vocal, were likely to be considered, but the Scientific Committee was not in a position to evaluate these.

In the Commission meeting that followed, these matters were, indeed, raised and, after much discussion the Commission finally agreed to a quota of 12 bowheads taken, or 18 struck, a markedly lower quota than the AEWC had hoped for. Even this was obtained only at the cost of some rather doubtful horse-trading on the part of the USA. To secure this limited agreement to their proposals they had had to agree to an enormous increase in the quota of sperm whales to be taken commercially by Japan and the USSR – from 763 to 6,444. Sperm whales were in no way an endangered species, but this reversal of previous American policy, in order to secure agreement to kill the undoubtedly endangered bowhead, shook those who had previously believed in the USA's commitment to whale conservation.

It was left to the Eskimos to divide the quota of 12 whales landed among the various communities, and considerable good-will was shown by them in so doing. Unfortunately, even in the initial season all did not go smoothly. The village of Barrow had been allocated three whales out of the quota, and, the ice conditions and weather being good, these were quickly taken. However, the Eskimos reported this catch as one bowhead and two '*inutuqs*'. *Inutuqs* were claimed by the Eskimos to be an entirely separate sort of whale, recognisable by its small size, fatter condition and different way of breeching (a whale is said to 'breech' when it jumps clear of the water). Such distinctions have eluded zoologists and it is difficult to see this claim as anything other than a subterfuge making use of the Eskimo custom of giving very different names to animals of the same species but of various age-classes, sexes or even behaving differently, in order to evade the quota restrictions. Barrow went on to take a fourth whale, but the authorities decided to ignore the violation.

64

Meanwhile, work had been set in train to get a better estimate of the bowhead population. Biologists, braving the worst of the North Slope weather, camped by the ice-edge to watch the whales go by on their migration. Hydrophones were suspended through holes cut in the ice to listen to the sounds of the whales, and spotter aircraft patrolled the leads, looking for the shadowy shapes of bowheads as they swam just beneath the surface. But whales are always difficult to spot, so whatever the number observed there was always the chance that some had been missed and the resulting estimates were too low. This provided a perfect opportunity for the Eskimos to claim that, because the populations were underestimated, the quotas should be raised, totally ignoring the fact that even were the populations twice what the Eskimos claimed them to be, there would still be no rational course other than total protection.

The bowhead population had, before the research efforts started, been estimated to number between 600 and 1,800 whales. The census of 1978 came up with a total of 2,264. For the next three seasons the estimated numbers were either close to the 1978 figure, or poor weather prevented very accurate counts being made. In 1981 an extensive aerial survey was conducted. This was financed by a consortium of oil companies concerned in developing the petroleum deposits of the Alaskan and Canadian North Slope, and with a strong interest in showing their environmental sympathies. Air counts, of course, record only those whales actually at the surface of the sea, but corrections for the numbers submerged were made from many hours of shore observation of the diving and surfacing behaviour of bowheads. The resulting estimates ranged from 3,199 to 4,427, with a mean of 3,842.

The following year, 1982, a very similar estimate of 3,857 (3,390–4,325) was made from ice-camps. In 1984, again from ice-camps, but with the additional aid of sophisticated sound-detection and location equipment, an estimate of 3,871 was made. Subsequent surveys and re-examination and analysis of previous counts now gives us a figure of around 4,400 as the best estimate for the western Arctic bowhead stock. This is perhaps 15 per cent of the pre-exploitation population.

Quotas for the Eskimo hunters have varied since they were first fixed in 1978. At the 1985 meeting of the IWC, the USA sought a quota of 35 strikes for the 1986 season. This was opposed by the Scientific Committee and, in the end, agreement was reached on a quota of 26 strikes a year for 1985, 1986 and 1987, provided that no more than 32 strikes occurred in any one year and that unused strikes could be carried over from year to year.

The arguments for and against aboriginal whaling on the depleted bowhead stock are complicated. The IWC set up a Technical Committee to examine Eskimo claims that the hunt was essential for nutritional and cultural reasons. Unsurprisingly, the Technical Committee concluded that Alaskan Eskimos had no special nutritional requirements and that any risk to the survival of the bowheads, which might be posed by the continuance of aboriginal whaling, could not be justified on nutritional grounds. On the other hand, it was accepted that the cessation of whaling would have a significant impact on the culture of the whaling communities.

The panel of experts set up to look into the cultural issues held that if whale hunting were abolished, the Eskimos would not starve; they would turn to a greater use of imported food, or would increase their harvests of seals, walrus or fish. But such alternative resources would not replace bowhead whales in the life of the Eskimos. Not only did they prefer the meat and '*muktuk*' (skin and blubber) of the bowheads to all other foods, but the activities of whaling served as a primary focal subsistence tradition.

Some people may find it hard to accept that the shooting of whales with modern bomb guns from *umiaks* equipped with outboard motors and CB radios has very much to do with Eskimo culture, particularly as no respect seems nowadays to be paid to the religious ceremony that once invested the whale hunt. Others hold that all cultures change and have done so throughout the course of human history; what is important is how the people concerned view these changes. In the case of bowhead whaling, the Eskimos have accepted the changes that have occurred and still attribute the highest cultural value to the hunt. This is the argument that has been accepted by the IWC, after some very pressing lobbying from the US. I find myself wondering if the US delegates would have argued so convincingly if the aboriginals concerned had been Siberians and the whales they wished to kill Californian grey whales on their northern migration.

For now, at any rate, the Eskimos have won and the bowheads have lost. But perhaps it is not a tragedy; the bowhead stock is believed to be large enough to withstand whaling at this level for many years to come and one might hope that if monitoring indicated that a radical reduction was occurring, the USA would have courage enough to ban the hunt. (If whaling really is essential to Eskimo culture, then the worst thing for both parties would be the extermination of the whales.) But whatever the arguments on population grounds, the cruelty of killing large whales by firing small bombs into their bodies remains. Contemplation of this must cause some disquiet to a good many people involved in making decisions concerning these whales.

The pygmy right whale

Before leaving the subject of right whales, there is one other member of the family that must be described. This is the pygmy right whale *(Caperea marginata)*. This strange whale is one of the least known cetaceans; in fact only 71 seem to be known. Prior to the 1960s, almost all our information came from stranded specimens. The pygmy right whale is found only in the

Fig. 3.7   The pygmy right whale, *Caperea marginata*.

Pygmy right whale, *Caperea marginata*, photographed at Plettenberg Bay, South Africa, in 1967. This is believed to be the only photograph of a living pygmy right whale.

Southern Hemisphere, in the waters between the tropics and the Antarctic. As its name implies, it is very small for a right whale, growing to only about 6.1 m (20 ft). It has small, rather narrow flippers and a small triangular dorsal fin, a feature not found in either of its larger cousins. There are 230–250 pairs of baleen plates, the longest being about 70 cm (27 in) in length.

The oddest feature of the pygmy right whale is that it is the cetacean with the greatest number of ribs – 17 pairs. These are peculiarly flattened on the underside, which led some workers to speculate that it was a deep diver that actually spent time lying on the bottom. Why a surface-feeder like a whalebone whale should lie on the bottom was not explained. In fact, further observations of living pygmy right whales, which became possible when living specimens swam into False Bay in South Africa on several occasions in the 1960s, indicated that they do not submerge for long periods, and presumably do not dive deeply.

When surfacing to breathe, the whole head of the pygmy right whale breaks the surface, in a manner very similar to that of the minke whale, the smallest and commonest of the rorquals. In consequence, it is likely that this species is often mistaken for the minke whale. The behaviour of this little whale at sea has been described as 'unspectacular', and this may be another reason why so little is known about it. It is of considerable interest to zoologists, as it is by far the smallest of the filter-feeding whales. Filter feeding is believed to put a premium on size, and a study of the energetics of the pygmy right whale could be fascinating.

67

Chapter 4
Humpbacks and Grey Whales

All whales are odd, but humpbacks and grey whales are among the oddest of the lot. Humpbacks, with their plaintive melodious songs and boisterous displays, have attracted many to become whale-watchers. Grey whales, with their predictable migration patterns and proximity to the vast population of coastal California, have done nearly as much, despite their rather more sinister soubriquet of 'devil-fish'.

Humpbacked whales

The humpback whale, *Megaptera novaeangliae*, is a stout, thick-bodied whale with enormously long flippers that may reach almost a third of the body length (Fig. 4.1). Although it is a member of the family Balaenopteridae, which includes the other whales of the genus *Balaenoptera* described in Chapter 2, it is sufficiently different to be classed in a genus of its own. *Megaptera* means 'long-winged' and refers to its overgrown flippers. The second part of its Latin name used at one time to be *nodosa*, which referred to the knobbles that adorn its short, broad snout, chin and the sides of the

Fig. 4.1   The humpback whale, *Megaptera novaeangliae*.

lower jaw. The great flippers are scalloped on their trailing edges, and this adds to the generally knobbly appearance of the whale. But the scientific names of animals are governed by a strict set of rules, of which the most pre-eminent is that of priority. When it was discovered that G. H. Borowski had, in 1781, described this whale as coming from New England (*novae-angliae*), the apter and less cumbersome *nodosa* had to be set aside.

Humpbacks generally grow to about 19 m (62 ft 4 in) in length, but, because of their very bulky body shape, they may weigh up to 48 tonnes. They are basically black in colour, although there are usually patches of white of varying size, most often on the belly, undersurface of the flippers and beneath the flukes. Colour patterns are sufficiently characteristic for them to be used to recognise individuals. The humpback's convenient habit of exposing the undersurface of its flukes just before it dives has allowed whale workers to build up files of photographs which can be used, for example, to follow the seasonal patterns of movement of these whales.

There are 270–400 pairs of black to olive-brown baleen plates, up to 42 cm (16 in) long. The throat grooves are quite unlike those of the other rorquals. There are only 14–24 very wide grooves, which reach at least to the navel. Humpbacks are almost always infested with whale-lice and barnacles in great numbers.

Humpback whales are cosmopolitan, being found in all the oceans and ranging from the tropics, where they breed mostly in the waters around island groups or continental margins, almost to the edges of the pack-ice. There appear to be three geographically isolated populations, one each in the North Pacific, North Atlantic and the Southern Ocean. The North Pacific stock spends the summer feeding in the Bering and Chukchi Seas, then migrates southwards along the western American coastline, before turning west to winter around Hawaii.

The North Atlantic population is composed of two groups. One, which

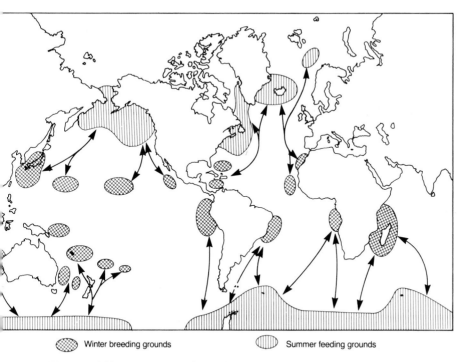

Winter breeding grounds   Summer feeding grounds

Fig. 4.2 Migration routes of humpback whales. These are nearly all north-south, from summer feeding grounds in high latitudes, to winter breeding areas in warmer water.

A humpback whale, *Megaptera novaeangliae*, rears its head clear of the water. The white patch beneath the chin is a mass of barnacles, infested with whale-lice.

summers to the east of Greenland and in the Barents Sea, migrates south along the European coast to the upper North Atlantic waters. The other group, from the south of Greenland, migrates south through the western North Atlantic to breed off Bermuda and the West Indies. The Southern Ocean population migrates north to the coasts of South America, South Africa, Australasia and various South Pacific islands (Fig. 4.2).

Humpback whales are slow swimmers, travelling at about 6–12 km/hr ($11-21\frac{1}{2}$ knots), and, because of their predictable, coast-hugging migrations and their tendency to congregate on both summer and winter grounds, they have been an easy prey to both aboriginal and modern commercial whalers. The early years of the great Antarctic whaling industry depended almost entirely on humpbacks, which could be found by the hundred and were very easy to kill. All stocks were radically reduced and some were almost certainly exterminated. Despite complete protection since 1964, there are few encouraging signs of recovery, as they seem to breed rather slowly. The world population of humpbacks today is of the order of 10,000, perhaps only 9 per cent of the pre-exploitation stock.

A humpback's tail has a pattern that can be used to identify individual whales.

Feeding

Two aspects of the humpback's natural history are of particular interest to me: its feeding and its singing. Humpbacks take a wide variety of foods, although they appear to restrict their feeding to their stay in cold waters. In the Southern Ocean they feed largely on the very abundant Antarctic krill, *Euphausia superba*, but in the Northern Hemisphere they may take, besides planktonic crustaceans, a wide variety of schooling fish, such as sardine, mackerel, anchovy and capelin. We have received this information mostly from the examination of the stomach contents of humpbacks that were killed in commercial whaling operations. Today, with total protection of humpbacks (apart from very small subsistence whaling operations in Greenland, the Caribbean and Tonga), it is less easy to make direct observations on what the whales eat. However, the very approachable nature of humpbacks, coupled with the great increase in the number of people interested in watching living whales, has meant that in the past few years much information on *how* the humpbacks feed has become available, helping to throw light on the feeding methods of other baleen whales.

Humpback whales have been observed to gather in areas where concentrations of planktonic organisms or small schooling fish occur. It is likely that the fish, which can swim, are attracted to the plankton. The locomotory abilities of plankton (by definition) are very limited. Some of the larger forms, like krill for example, swim actively enough to form dense patches or swarms, but in other cases dense concentrations of plankton are formed where two water-masses meet.

Whatever the reason for the concentration of the food organisms, the whales seek these out and feed there selectively. Like fin whales, humpbacks often harvest these concentrations by lunge feeding.

The technique is very similar to that of the fin whale. The humpback swims towards a patch of food, often shallowly submerged, but at other times almost directly below, and then lunges up, at the same time turning to bring the gape of the huge jaws around the prey. During this turn at the end of the lunge, the whale's huge flipper may sweep out of the water and down again with a great splash. Whether this is merely part of the physical contortions involved in manoeuvring, or whether the flipper plays a part in concentrating the prey, we do not know. The leading edge of the humpback's flipper is a brilliant white and this might serve, as it crashes down into the water, to scare schooling fish or plankton away and towards the mouth. The huge size of the flippers of the humpback, in comparison with those of all other rorquals, suggests that they have some particular role to play, but my guess is that they are used in social signalling rather than in feeding.

Frequently more than one whale is present during lunge feeding and their movements appear to be synchronised. Several humpbacks have been described as making a series of long vertical dives to herd a shoal of herring together before beginning lunge feeding. It is difficult not to see such actions as co-operative behaviour among the whales, although some biologists believe that there is a good deal of competition in such feeding. The two strategies are not necessarily mutually exclusive, of course.

Another way humpbacks get food is called 'flick feeding'. In flick feeding the whale swims in what appears to be a normal dive just below the surface, but then suddenly the tail is flexed forward, checking the forward movement of the whale and generating a wave through which the whale then swims with its mouth open. The wave is believed to concentrate small food items, such as euphausiids, though precisely how this works is far from clear.

But the most remarkable humpback feeding method of all is bubble-netting. This was first recorded by a Norwegian whaler in the Arctic Ocean, but nearly 40 years were to pass until Charles and Virginia Jurasz noted it again in 1968 in south-eastern Alaska. When humpbacks encounter swarms of krill at or near the surface one, or sometimes several, whales will swim in a upward spiral from below the swarm, at the same time releasing a stream of air from its blowhole. The rising bubbles appear to create a cylindrical curtain which has the effect of corralling the euphausiids, which then swim towards the centre, or, in other words, away from the water disturbed by the bubble-net. The whale, together with its companions, then rises to the surface to feed on the resulting concentration.

As many as eight humpbacks at a time have been seen to feed at one bubble-net, although, as in lunge feeding, there is doubt as to whether this really represents co-operative behaviour. Two humpbacks have been seen to construct a bubble-net, only to have its contents raided by two others that surfaced within it.

A strange aspect of the many observations on humpback feeding is that different animals seem to have different feeding methods, which they use fairly consistently. A group of observers from the Cetacean Research Unit, at Gloucester in Massachusetts, watched humpbacks feeding. Seventeen of these animals had distinctive feeding styles. Eleven of them would create a disturbance at the surface of the water by lob-tailing (smacking the water surface with their tail flukes) or breaching (jumping clear, or nearly clear, of the water surface) before following this up with a series of feeding lunges. Four used bubble-nets (unusual in the Atlantic) and two were vertical lunge feeders. One of the whales watched was a juvenile. Born in 1983, it had developed its own distinctive feeding pattern by 1985, although its mother did not have a distinctive pattern herself.

Observations such as these are made possible by the opportunity to recognise individual humpbacks from their characteristic patterns and general appearance and also by the very long hours of patient observation that whale watchers are prepared to put in. Gradually, by their efforts, we know a little more about the biology and individuality of the whales.

The song of the humpback

In 1970 a long-playing record was put on the market that was unlike anything that had come before. This was *Songs of the Humpback Whale* (CRM Records, California, SWR-II). These eerie, haunting notes of the hump-back whale, brought to a wide public through this record, made a deep impression on many listeners. Very few people had heard these sounds before and even fewer had commented on them. Humpback vocalisations are audible through the hull of a wooden ship and sailors in the past, lying becalmed on a tropical sea, must have heard these strange plaintive sounds. No doubt they thought them some supernatural product – they are certainly haunting enough to be so regarded. Perhaps they were the origin of the songs of the sirens that failed to lure Odysseus, lashed to the mast, to his doom.

One nineteenth-century American whale captain, C. Nordhoff, who wrote *Whaling and Fishing*, was well aware that a humpback might sometimes get beneath a vessel and 'utter the most doleful groans, interspersed with a gurgling sound such as a drowning man may be supposed to make'. He went on to tell a yarn of how his crew, unfamiliar with such sounds, supposed the ship to be haunted and deserted to a neighbouring ship. Later, in 1883, a log book entry from the bark *Gay Head* records catching a 'singer' off the west coast of Central America on 15 August. The date and location make it likely that this refers to a humpback whale, although the term 'singer' does not seem to have been in wide use in the American whale fishery.

The vocal performances of whales fell into obscurity until, during World War 2, the US Navy set up underwater listening stations to detect the movements of enemy submarines. These SOFAR (Sound Fixing and Ranging) stations found that the sea was a comparatively noisy environment. They had considerable difficulty in separating the sounds of intruding submarines from those of biological origin. For obvious reasons, little information about what was heard was made public until after the war had ended. It was not till 1952 that an article in the *Journal of the Acoustical Society of America*, by O. W. Schreiber, described sounds in the low audio-frequency region, with a rather musical quality, that had been picked up by the SOFAR station at Kaneohe Bay in Hawaii. Schreiber noted that there was a marked seasonal variation in the occurrence of the sounds, and that this corresponded with the seasonal variation of the presence of whales in the area; from this he concluded that the sounds were produced by the whales.

The pioneers in the study of the song of the humpback whale were Roger Payne and Howard and Lois Wynn. Roger Payne had gone to Bermuda with his wife, Katy, in 1967 to study humpbacks. While there he met Frank Watlington, an engineer with the Columbia University Geophysical Field Station at St David's. Frank Watlington, in the course of his studies of sound propagation in the ocean, had recorded a number of strange marine sounds, which he later decided were made by humpbacks. Payne listened, fascinated, to these recordings and realised that the various sounds were not simply a random collection of calls, but had a definite sequence and order, and he decided that they merited being called songs. Animal songs, of course, were familiar enough in birds or grasshoppers, but no one had previously recorded them in whales and Roger Payne decided to devote his research to their study.

A written description of the song of the humpback whale is, of course, no substitute for listening to the real thing. To know that the sounds are mostly in the frequency range 40–5,000 Hz and that there are six basic types of song, each of which can be sung in one of several variations, gives not the slightest idea of the haunting beauty of the humpback's repertoire. But if I am to say anything at all about this subject, which fascinates me so much, I must use the words available. (It is at this stage that the wise reader will go out and buy or borrow the record.)

Most of the jargon for describing songs in the animal kingdom comes from the study of birds, for these are the singers we are most familiar with. In bird terms, then, the humpback sings a true song, consisting of an ordered sequence of themes comprising motifs and phrases made up of syllables. The syllables are the simplest part of the song and they have been given more or less descriptive names: moans, cries, chirps, yups, oos, surface rachets and snores. Of these seven sounds only chirps and surface rachets are missing from some calls; the others are always present in some form, although they may be combined in different ways. Of course, the researchers have subdivided these basic sounds and talk of wos, foos, mups, ups and so on, as well as the others listed above, to make up about 20 syllables.

The duration can be very variable, mostly between six and 35 minutes, although one humpback recorded by the Wynns sang continuously for 24 hours and was still singing when they had to leave. Within a singing session each song is clearly defined, with a recognisable beginning and end, often marked by the 'surface rachet' syllable. If a song is interrupted, by breathing, for instance, it is taken up again where it left off. The motifs in the phrases, and the phrases in the themes can be repeated any number of times, and this accounts for the variation in length of the songs. Further individuality is interjected into the song by the inclusion of odd chirps and cries in the main composition.

While songs are individually recognisable, all the whales in one region sing one basic song. So far three dialects have been recognised, one in the North Atlantic (the West Indies and the Cape Verde Islands), one in the North Pacific (Hawaii and to the west of Mexico) and one around Tonga in the South Pacific. Perhaps if other humpback breeding areas, such as those that once existed off Madagascar or the west coast of Australia, were visited, it might be found that there were different dialects in use there, although, if we are right in concluding that there is only one breeding stock of humpbacks in the Southern Ocean, it may be that there is only one dialect. Songs change gradually over the course of several seasons, and the changes are followed by all the whales in that area.

It is not possible to say exactly how far off these songs are audible to other whales, but the general range of the song is probably detectable at least 31 km (19.2 miles) away, with the very low-frequency snores and moans having a range of at least 185 km (115 miles). The repetition of phrases and motifs may help to make the song interpretable to other whales at extreme ranges.

It is not certain how the humpbacks produce their songs. The usual mammalian way of vocalising is to pass air from the lungs through the larynx over the vocal cords, which can be tautened by very precise muscular control to regulate the frequency produced by their vibration. While this is being done, the mouth is held open and the quality of the sound produced is affected by the resonance of the buccal or nasal cavities, or perhaps of specially developed organs, like the trunk of an elephant seal or the bony voice-box of a howler monkey. Whales do not have vocal cords and, as they sing while submerged, they cannot afford to release large amounts of air to the exterior. Nevertheless, it seems likely that the songs are produced by the movement of air setting in vibration some specialised tissue, perhaps part of the larynx. It may even be that more than one site of sound production is involved. The low and very resonant notes produced probably reflect the huge scale of the whale's vocal apparatus.

Whatever its origin, the humpback's song is clearly an important part of its behaviour. Its complicated form means that, potentially, it can convey a lot of information. What then is its function? We have no clear answer to this, but we do have a good many clues that give us some strong indications. Firstly, the songs are sung on the breeding grounds. Only exceptionally is a humpback heard singing on the summer feeding grounds. Secondly, it is only mature males that sing, a fact that has been determined by direct observation by divers in the water with the whales, and also by taking biopsy samples of the skin of singing whales and examining these for the telltale chromatin bodies that are always found in cells from females. These were never present in singing whales, and it is safe to assume that all the singers are males.

If we look at the behaviour of the whales on their wintering grounds we find a loose social structure. The strongest bond is between the mother whales and their calves, but, whether with a calf or not, a female humpback forms the nucleus of any social formation that develops in the area. There is good evidence that individual whales return to the same site year after year. The males arrive in shallow coastal waters with a depth of about 20–40 m (66–131 ft), take up position and start singing their songs. A female cruising by will attract a singing male, who joins her, stops singing, and becomes her 'principal escort'. If accepted, the principal escort swims alongside the female, in close contact, perhaps gently touching flippers, or rolling or diving together. It has been suggested that during such dives mating may take place.

If males are abundant in the area the pair are likely soon to be joined by another male and a good deal of aggressive jostling and barging for position near the female may then take place. The rival males will lunge at each other with their snouts, or thrash with their tail flukes and emit bubble streams from their blowholes. Although they do not sing while fighting, they produce a variety of vocalisations, including song segments out of context, and the noise soon attracts other males to the scene. Eventually the group may number as many as 15 animals, and then one of the newcomers displaces the principal escort and takes his place. Fidelity to a single male thus seems unlikely. Whales that join the group are frequently singing, but this stops while they are in the group, although they generally resume singing if they leave it.

It thus seems very probable that, as in many birds, the whale's song is associated with reproduction. Peter Tyack has suggested that the song may act as a spacing mechanism for sexually active males, which thus advertise their availability to receptive females. This may well be so, but it is not obvious to me why such a complicated signalling device is needed for this. What role is played by the seasonal and longer-term change in the song? Why do only humpbacks, but none of the other baleen whales, have songs? Comparisons made with lekking behaviour in game birds are perhaps a little premature at this stage of our understanding of the humpback's love-life. Like so much else of the natural history of whales, the haunting melody of the humpback's song remains shrouded in mystery.

The grey whale

The grey whale is probably the most primitive of the living baleen whales. Some people have thought there are direct links between the Oligocene cetotheres and present-day grey whales, but this is a misconception. In fact, we can trace the origin of grey whales back no further than the Pleistocene era. This is a pity, for in appearance the grey whale seems to be a very suitable candidate for a missing link between the right whales and the rorquals, and missing links are eagerly sought after in popular science.

Fig. 4.3   The grey whale, *Eschrictius robustus*.

Grey whales, *Eschrictius robustus* (Fig. 4.3), average about 12 m (40 ft) in length, but can be as long as 15 m (50 ft). The body is more slender than that of the right whales, but stouter than that of most of the other rorquals (although not the humpback). The head is smaller in proportion to the rest of the body than in the other baleen whales. On the throat there are between two and four ventral grooves, about 1.5 m (5 ft) long. There is no dorsal fin, but the dorsal ridge of the hind third of the body is surmounted by a series of low humps or crenellations, numbering between six and a dozen. (This is similar to the condition found in the sperm whale.)

The body, as the name suggests, is grey with whitish mottlings. It usually supports thriving colonies of barnacles and whale-lice, and it may be that the scars caused by these organisms are responsible for the mottled appearance. Hanging from the upper jaw are between 140 and 180 pairs of rather coarse whitish baleen plates, up to 40 cm (16 in) in length, with very long, thick bristles. If one gets the chance to examine the head of a grey whale closely, it is possible to see that it has many more tactile hairs scattered about its snout than the other baleen whales.

Grey whales are the most coast-hugging of the baleen whales, and it is probably this feature which accounts for the fact that they are now reduced to only a fraction of their original abundance. They are a Northern Hemisphere species and used to occur in both the North Atlantic and the North Pacific. The species was, in fact, originally named from subfossil material found in Sweden in 1859. This was followed by other discoveries in the Low Countries and East Anglia and it was soon found that these were identical to bones of the Californian grey whale, described by Cope in 1868, which was then the subject of a vigorous fishery.

The history of the grey whale in the North Atlantic is a curious one. Towards the end of the ninth century, Alfred the Great, King of Wessex, summoned a Norwegian called Othere to his court to tell him what he knew about Norway and its resources. Othere described, among other remarkable things, how he and five companions had 'in the course of two days slain sixty whales, each forty-eight ells long, the largest fifty'. The value of the Norwegian ell at that time is not certain, but it seems likely that one ell was equivalent to one or two English feet (30.5 or 61 cm). The lower value would fit the grey whale rather well, and the known behaviour of this species and its vulnerability to attack make Othere's rather startling claim seem somewhat less unbelievable than it appears at first hearing.

Eight hundred years were to elapse before the next reference to a

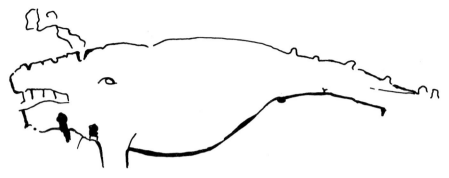

Fig. 4.4   Jon Gudmundsson's picture of the '*sandloegja*', which may have been a grey whale from Iceland in past historical times.

European grey whale. Early in the seventeenth century, in Iceland, Jon Gudmundsson the Learned wrote a treatise which he called 'The Book of Whales'. Jon's accounts of whales and other sea creatures are often fantastic and the illustrations hard to reconcile with any known species, but one of the whales he describes, the '*sandloegja*', has characteristics that link it to the grey whale. Jon's picture (Fig. 4.4) shows a rather bulky whale, with baleen plates hanging from its upper jaw, no dorsal fin or ventral grooves (there are a couple of blots over the throat, so we cannot be sure that some short grooves were not originally depicted), but with a series of six low knobs irregularly spaced on the hind part of the body. In his description, Jon relates the following: 'Sandloegja. Good eating. It has white baleen plates, which project from the upper jaw instead of teeth, as in all other baleen whales ... It is very tenacious of life and can come on land to lie as seals to rest the whole day'.

The only whale besides the grey whale which has white baleen is the minke whale, and this has a conspicuous dorsal fin, which Jon would almost certainly have included in his drawing. The six knobs in the tail region might well represent the crenellations of a grey whale; the only other whales they would fit would be a sperm whale, which has teeth and not baleen, or a humpback, with black baleen and a fin-like prominence on the back. Jon's comment about resting on the sand ('*sandloegja*' means 'sand-lier') fits the observed behaviour of grey whales in lower California, where they will lie on the bottom in very shallow water, although they do not, of course, crawl out as seals do.

The next reference was from the other side of the Atlantic. Early in the eighteenth century the New Englanders were hunting a whale they called the 'scrag whale' along their coasts. This is how Peter Dudley described the scrag whale in the Philosophical Transactions of the Royal Society of London in 1725:

'The Scrag Whale is a near kin to the Fin-Back, but, instead of a Fin upon his Back, the Ridge of the after part of his back is scragged with half a Dozen Knobs or Nuckles; he is nearest the Right Whale in Figure and for Quantity of Oil, his Bone is white but won't split.'

This was undoubtedly a grey whale, but it is the last we hear of that species from New England, or, indeed, from the Altantic. Evidently the last few survivors were boiled down for oil by the hunters.

Hunting the grey whale

The grey whale fared a little better in the Pacific. Originally there were two stocks, one summering in the Sea of Okhotsk and migrating down the western side of the Sea of Japan to breed in the bays of southern Korea; the other summering in the western part of the Bering Sea and migrating down the west coast of North America to breed in the lagoons of Mexico. Its liking for shoal water made it conveniently accessible to primitive hunters in Japan, Kamchatka and Siberia on one side of the Pacific, and from Alaska all the way down the coast to Baja California on the other. However, as in the case of the bowheads, this pressure from subsistence hunters seems to have been sustained by the stocks. It was not until the mid-nineteenth century, when commercial whaling for this species started off the Californian coast and in the Mexican lagoons, that the grey whale came under serious threat.

Most of what we know about the history of this period is derived from the writings of an American whaler from Maine, Captain Charles Scammon. Scammon wrote a classic work: *The Marine Mammals of the Northwestern Coast of North America, Together with an Account of the American Whale-Fishery*, which he published in 1874 after he had given up whaling. Scammon's book provides a mine of information, which is still useful today, about the whales and seals of that area, and the methods used to hunt them. But, besides being a talented author, Scammon was also a very successful whaler. In 1855 he had the good fortune to discover a lagoon, now officially known as Laguna Ojo de Liebre, but almost always referred to in English as Scammon's Lagoon, where the grey whales bred and where it was easy, in the sheltered and restricted waters, for the whalers to kill them.

Grey whales enter the lagoons to give birth to their young. The shelter, and perhaps the higher temperatures of the lagoons compared with the open ocean, are probably significant for the newborn calf. Scammon describes how, in December to March, the pregnant females would collect together at the most remote extremities of the lagoons, huddled so thickly that it was difficult for a boat to cross the waters without coming into contact with them. 'Repeated instances have been known of their getting aground and lying for several hours in but two or three feet of water, without apparent injury from resting heavily on the sandy bottom, until the rising tide floated them'. This remark reminds us of Jon Gudmundsson's comment on the *sandloegja* coming on land to rest the whole day. Of course, in such a situation the whalers killed them by the hundred. Had it not been for the decline in demand for whale products just before the turn of the century, when petroleum was taking the place of whale oil, it is likely that the grey whale would have been as surely exterminated in the eastern Pacific as it had been in the Atlantic.

But the grey whale was not entirely passive in the face of these assaults.

Scammon remarked that the grey whale was the most dangerous of all whales to attack, and that compared with the number of ships which were formerly engaged in their capture, more casualties had occurred than in any other branch of whaling. He goes on to recall how, in 1856, when hunting in Magdalena Bay, two boats were entirely destroyed, the others were staved 15 times, and, out of 18 men who manned them, six were badly jarred, one had both legs broken, another had three ribs fractured, and still another was so badly injured internally that he was unable to perform any duties during the rest of the voyage. And all this happened before a single whale was captured! It was small wonder that the American whalers called them 'devil fish'. *Quel animal affreux; quand l'on attaque, il se défend!*

The assaults on the grey whale came later on the other side of the Pacific. It was not until modern steam whaling was introduced there, in the closing years of the nineteenth century, that the catch rose much above subsistence levels. The western stock, which may have been smaller than that on the eastern side of the Pacific, was reduced to a tiny remnant, and the Korean grey whale is currently endangered, if not already extinct. The International Whaling Commission prohibited a commercial take of grey whales in 1946 (quite a lot of whaling had gone on off the Californian coast between the mid-1920s and the end of World War 2), but the Soviet Union was permitted to take a small number of greys for the aboriginal people living on the Chukotskiy Peninsula. The whales are taken by modern whale catcher boats and then handed over to the native people. However, there are suspicions that this catch is more to feed commercially farmed mink than aboriginal people, and several conservation groups have protested. Greenpeace at one time even went so far as to stage a small-scale invasion to draw attention to this harvest.

The Siberian catch of grey whales probably comes from the stock that breeds in Mexico, but it is a small harvest and, since the protection order of 1946, there has been an encouraging increase in the number of grey whales. Because of their coastal migration paths, it is a comparatively easy task to count them as they go by. Migrating grey whales are slow swimmers, making about 8 km/hr (4.5 knots), although they can speed up to 20 km/hr (11 knots) when stressed. They swim steadily, surfacing to blow three to five times every three to four minutes. The tail flukes are raised out of the water on the last blow before the whale dives.

Such accessible and impressive creatures have become a tourist attraction in California. It has been calculated that about a million visitors a year gather on Point Loma above San Diego Bay to watch the grey whales go by a few hundred metres off the rocks. Many small launches take whale-watchers to meet the whales further out to sea, but US federal law requires that the whales shall not be harassed, so the boats have to stand off a good distance. With luck, however, a whale may swim up to a boat that is lying still, giving the passengers a sight they will never forget.

The prohibition on hunting grey whales has been so successful that the current population, which is estimated to be about 18,000, seems to be not far off its pre-exploitation level, which was probably around 20,000 in the North Pacific. The grey whale is, in fact, the only whale which, having suffered near-extinction, has shown itself to be resilient enough to make a substantial recovery. The reason for this is not obvious. From what we know of its breeding habits, it does not appear to be especially fecund. Females reach puberty at about eight years, and bear a single calf in alternate years – the common pattern in baleen whales. It would be good to know that right whales and bowheads could do as well.

Feeding

Biologically, perhaps the most interesting feature of the grey whale is its method of feeding. We have seen how, in the rest of the baleen whales, the whole structure of the head has been modified to create a complex apparatus for harvesting plankton from the sea. But the grey whale does not feed like that. Until recently, when enterprising people began to dive with grey whales ('devil fish', indeed!), what we knew about their feeding habits had been learnt by deduction. Grey whales do most of their feeding in their Arctic summering grounds, while most of the commercial catch was taken while they were on migration or on their winter, breeding, grounds. A consequence of this was that most of the reports of stomach contents provided little information on diet. It was observed, however, that the baleen of the right side of the mouth was usually more worn than that on the left, and that the stomach frequently contained quantities of sand or gravel. Today, however, we have direct evidence from underwater observation of grey whales feeding in the wild and even in captivity.

Grey whales are predominantly, although not exclusively, bottom feeders. They feed mostly in shallow water between 5 and 100 m deep (15–330 ft), which accounts for their coastal distribution, taking those organisms that live either on the bottom sediment (epifaunal organisms) or within it (infaunal organisms). The sea bottom in a productive ocean supports a rich community of invertebrates, which live on the detritus falling from the upper layers where the phytoplankton synthesises organic material from carbon dioxide, mineral salts and energy from the sun. The food chains in the sediment are complex, but the prominent consumers are more or less sedentary shrimp-like crustaceans, the gammarid amphipods, together with various polychaete worms and molluscs. Although the grey whale takes a wide variety of invertebrates when feeding on the bottom, it is the gammarid amphipod, *Ampelisca macrocephala*, that is probably its commonest food.

When feeding, the whale swims towards the bottom, turns on one side, usually the right (this accounts for the uneven wear on the baleen), and ploughs its snout through the top layer of sediment. This stirs up the sediment, together with those creatures on the surface of it and in the top few centimetres. The whale sucks the disturbed water and sediment, together with its contained organisms, into its mouth, and then filters off the excess water and mud particles in the conventional mysticete fashion. It has been suggested that the whale can select the amphipods by timing the stage at which it takes the turbid water into its mouth – once disturbed, the amphipods tend to remain swimming while the less buoyant, or less motile, organisms sink more quickly. The very stiff, coarse and relatively short baleen plates, with their stout bristles, are well adapted for this abrading work.

Where grey whales have been feeding in this way, the sea bottom is marked with the furrows caused by their ploughing. It has been suggested that this can actually increase the productivity of the feeding area, although it is hard to see how disturbing the sediment would have this effect.

Another method used by the grey whale is termed 'suction feeding'. A young female, feeding on burrowing ghost shrimp (*Callianassa californiensis*) in Grice Bay, British Columbia, was watched and photographed by scuba divers. Instead of ploughing steadily through the sediment, the whale turned on her side and applied her lip against the bottom while maintaining her station using her flippers and tail flukes. She then sucked the sediment into her mouth, presumably by using her very muscular tongue, and filtered off the shrimps. Having exhausted one area, the whale would move on to another, leaving a series of shallow conical pits on the seafloor to mark her feeding places.

Although there is no substitute for observation in the wild, we do have some interesting information from a young female grey whale, named Gigi, that was held captive for a time (while recovering from an injury) at Sea World, San Diego, in 1971. Gigi was fed on squid (which were of good nutritional quality and easier to obtain in San Diego than gammarid amphipods or ghost shrimps). The dead squid (900 kg or 1,980 lb each day!) were tipped into her pool and lay on the bottom. Gigi would turn over on her side, usually, but not always, the right side, and seemed, by

A young female grey whale, *Eschrictius robustus*, in captivity.

tongue movements, to increase the volume of her mouth so that the lower lip was depressed and water and food flowed in by suction. While feeding, the throat region was slightly distended and the grooves expanded. The actual suction seemed to occur in pulses and sometimes a squid would reappear out of her mouth after having been sucked in.

Grey whales have been seen to take fronds of kelp into their mouths, and one Russian writer has suggested that they are to some extent vegetarian, although this would be a very exceptional feature in a whale. It seems more likely that what the whales are doing is scraping, or sucking, from the surface of the fronds of the kelp, the multitudinous tiny animals, such as hydroids, molluscs and tunicates, that use kelp as a substrate. Close observation of grey whales with kelp in their mouths shows that the kelp is later released undamaged. There is no indication that they ever deliberately swallow any of it, although fragments have been reported in stomachs.

Besides these less usual feeding methods, grey whales can feed on plankton or shoaling fish in the water column itself, like the other baleen whales. They have been seen criss-crossing through a dense school of small fish, perhaps anchovies, off San Diego; and one that was cast up on the beach at Grays Harbour, Washington, had several gallons of rainbow smelt in its gullet. Grey whales have also been observed to feed on swarms of mysids (another form of shrimp-like crustacean), which were swimming just above a rocky bottom. One can perhaps see this as an intermediate stage between sediment feeding and the more conventional mysticete style of grazing plankton.

Does the grey whale give us any hint of how feeding using a baleen filter developed? To sift plankton from the water column requires a fully developed filtering system, such as we see in the modern mysticetes. But we can suppose that an ancestral mysticete might have grubbed about on the bottom for its food, taking the larger organisms, crabs or sea cucumbers, for example, that could have been caught with a conventional mammalian mouth. In these circustances, teeth might have been of little value. They would wear against gravelly sediments, and perhaps they became reduced and non-functional as they are in so many of the toothed whales. Indeed, they may have been reduced already at this stage; the nature of the tooth germs in foetal mysticetes suggests a very simple dentition. Palatal ridges might then have taken the place of teeth, with the advantage that, though these would quickly wear, they could be replaced by growth from the dermal papillae that underlie them. Development of these palatal ridges might have been an advantage in sorting food items within the mouth, and further development might have made it possible for smaller and smaller invertebrate prey to be captured, leading eventually to the filtering mechanism we see in the modern grey whale.

This is, of course, speculation. Alas, the zoologist never really finds missing links to bridge the evolutionary gaps. All the extant forms available to us (and, equally, all the fossils) represent forms that have been modified to make the best of their environment as they find (or found) it. But of all the mysticetes, I think the grey whale is the nearest to the ancestral stock.

Chapter 5
Sperm Whales – Divers in the Deeps

The sperm whale is unlike any other whale, or, indeed, any other animal. Dale Rice, an American whale biologist, noted that if there were not already such an animal, one might say it was impossible! The sperm whale is a toothed whale, but like the baleen whales it has a huge head – about a third of the bulk of the animal – which, instead of being equipped with baleen, is filled with a strange liquid wax, or spermaceti. More than any other cetacean (apart from the unknown species that swallowed Jonah), the sperm whale has achieved literary distinction in the character of Moby Dick, the immense white whale that so plagued Herman Melville's Captain Ahab. *Moby Dick* should be required reading for all interested in whales. Besides being an author, Melville had, in his youth, shipped as a whaler, and there is a great deal of accurate observation about whales and their habits, as well as the wealth of allegory and drama that has made *Moby Dick* one of the most important works of American literature.

Fig. 5.1   The sperm whale, *Physeter catodon.*

In appearance the sperm whale, *Physeter catodon*, is unmistakable (Fig. 5.1). The huge head makes up 25–33 per cent of the body length. The proportion is considerably greater in males than in females (which are much smaller than males), and contributes to the very different appearance of the sexes. The single S-shaped blowhole is situated at the tip of the snout, well to the left of the midline. When a sperm whale surfaces to blow, the blunt square end to the head, and the position of the origin of the bushy spout, ejected at an angle of about 45°, make it one of the most easily recognised whales, even at a considerable distance.

Beneath the head is a narrow, rod-like lower jaw, armed with 18 to 28 pairs of massive conical teeth. These become much blunted in older animals, but how they become so abraded is hard to discover, for there are no opposing teeth in the upper jaw, the mandibular teeth simply fitting into fibrous sockets in the gum above. This statement is not *quite* true, however. Although there are no visible teeth to be seen in the upper jaw,

84

A sperm whale, *Physeter catodon*, raises its snout above the water to blow. The position of the blowhole, on the left side of the tip of the snout, is unique in whales.

often there are rudimentary teeth to be found concealed beneath the gum. The posterior halves of the lower jaw diverge to make their articulation with posterior halves of the lower jaw diverge to make their articulation with the skull, while leaving a wide gape through which the sperm whale can swallow its food. The angle of the jaw tilts up a little, giving the sperm whale a somewhat benign expression. Above and behind the angle of the jaw, and slightly below the middle of the side of the head, is the eye. This appears tiny in relation to the size of the whale, but it is, in fact, quite large enough to be fully functional and sperm whales appear to have good eyesight.

Behind the head, the sperm whale is less extraordinary. There are a pair of rather small, rounded flippers about a third of the way along the body, or, in other words, just behind the head. The tail flukes are broad and massive; their rather simple mechanical function as the organs of propulsion does not allow much variety of design in any whale. There is a sort of hump, often referred to as a dorsal fin, at the beginning of the hind third of the body, and following this the upper surface of the tailstock has a scalloped appearance, resembling the crenellations that occur in the grey whale.

The skin over the trunk of the sperm whale is irregularly corrugated, giving it a shrivelled appearance, although this is less noticeable in fat animals on their feeding grounds. In colour the sperm whale is a dark slaty-grey or brown, paler on the belly where white patches often occur. The insides of the mouth and lips are white but the tongue, when visible, is bright red. White sperm whales do appear from time to time – Moby Dick was not unique – but they are rare. Old sperm whales appear to become paler and certainly acquire a great many pale scars, often the marks of the suckers of the squid on which they feed, but in the case of males the scars may also result from sexual fighting.

Male sperm whales may reach 18.5 m (60 ft) in length, although few large ones now exist and animals longer than 45 m (50 ft) are rare. Females are much smaller, reaching only 13 m (43 ft). Males weigh between 32 and 45 tonnes; females about 16 tonnes. When we encounter

A sperm whale, *Physeter catodon*, on the flensing platform. The single nostril, on the left of the tip of the snout, is clearly shown.

The lower jaw of the sperm whale is armed with about 18–28 pairs of stout conical teeth.

such large discrepancies in the size of male and female mammals, it usually tells us something about their reproductive patterns. If, as in the baleen whales, the females are about the same size as the males, or a little larger, we generally find that there is no special social structure associated with breeding, and mating takes place on a one-to-one basis. Where males are noticeably larger than the females, as for example, in red deer or fur seals, we find that at the breeding season there is a good deal of challenging and actual fighting between the males, to try to accumulate a group of females, usually called a harem, to which the successful male has exclusive access to pass on his genes.

We have fewer opportunities to observe the mating rituals of sperm whales than those of fur seals or red deer, but what we can observe bears out the hypothesis that the size discrepancy in sperm whales is related to a highly polygamous breeding system.

Although sperm whales are cosmopolitan, being found in all the oceans of the world, they have a very uneven distribution (Fig. 5.2).

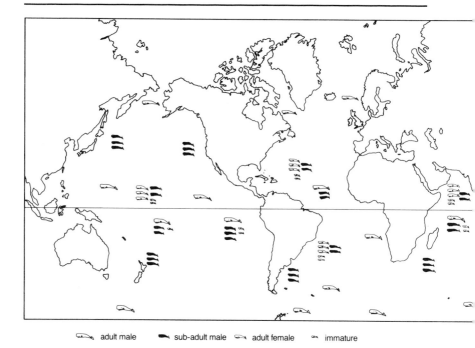

adult male   sub-adult male   adult female   immature

Fig. 5.2 The distribution and social structure of sperm whales. Only adult bulls enter the cold waters of high latitudes. Females and immatures are found in warmer waters. The groups with females are joined by adult males for short periods for mating. Herd sizes are variable, but average about 20 whales. (Based on a diagram in Peter Evans's *The Natural History of Whales*.)

Females and young are restricted to warm tropical waters, while adult males are also found in high polar seas. These polar bulls are often described as solitary, but may travel together in loosely associated groups.

In warmer waters the social system is more defined. The basic social unit is the nursery school. This is a mixed-sex group with a preponderance of females, which usually make up about three-quarters of the group, the males that are present being sexually immature. These schools number between 20 and 40 whales, which move about in a relatively tight formation and whose composition seems to be rather stable. Observations on tagged whales show that some females remain together for at least a decade.

During the breeding season, around December in the southern stocks and six months later in the northern ones, a nursery school is joined by a mature male. This bull becomes the 'schoolmaster' and he defends his position with the females against other bulls. During these contests, head-butting is the most frequently used means of attack, but the formidable teeth may also be used to swipe an opponent, and broken jaws can result.

Although observations on the frequency and incidence of mating in sperm whale schools are scarcely possible, it seems reasonably certain that the schoolmaster, while he maintains his position in the school, has exclusive access to the females that come into season. His presence within the group excludes other males, and thus the schoolmaster has a better than average chance to pass on his genes to the next generation; in other words, he has achieved a selective advantage. As larger bulls are presumably better able to defeat their opponents in the butting contests, while no such advantage will accrue to the females, we can see how the sexual dimorphism in size has evolved in the sperm whale.

Birth follows about 15 or 16 months after mating. The newborn calf of either sex is about 3.5–4.5 m(11 ft 6 in–14 ft 9 in) long and weighs about a tonne. The calves are suckled by their mothers in the nursery schools for at least two years, with an indication that the older mothers suckle their calves longest. Some 20 kg (44 lb) of milk, with a fat content of about 33 per cent, is supplied by the mother each day. Even after functional weaning, that is, when the young whale has become nutritionally independent, there is evidence (from traces of lactose, a sugar contained in milk found in the stomachs of whales killed by hunters) that some young whales may continue to suck for several more years, an activity which, perhaps, has a social function in extending the mother-young bond and also the integrity of the group.

Although born the same size, young male sperm whales soon begin to outgrow females of the same age. As the young males mature, they leave the nursery schools to form bachelor groups. Groups of small bachelors tend to stay in the same warm waters as the females, but, as they grow still larger, they later increase their wanderings to cooler seas. Only the largest males move into high latitudes, and, though these may travel as groups, they are often reported as solitary animals.

The spermaceti organ

We have seen that one of the functions of the huge head of the sperm whale is to be used as a ram to drive off opponents. But this function alone would not call for the complicated arrangements that make up the internal structure of this organ. Besides the usual features that one would expect to find in the head of a mammal, the sperm whale's head contains two massive structures, the 'case' and the 'junk', together known as the spermaceti organ. These structures make up the bulk of the head and are packed with a liquid wax called spermaceti. The function of this strange head and the spermaceti within it has long been a puzzle to workers on whales.

Some very clever deductive work by a British biologist, Malcolm Clarke, has shown that a main function of this part of the head, and possibly its principal function, is to act as a hydrostatic organ, controlling the buoyancy of the whale when it dives and ascends. The sperm whale is a prodigious diver, having been observed on sonar to descend below 1,200 m (3,936 ft). Even deeper descents have been deduced from the discovery of fresh specimens of bottom-dwelling sharks in the stomach of a sperm whale which was shot in an area where the depth of water was in excess of 3,200 m

(10,400 ft). These are record figures, and by no means all dives are as deep as these. Probably the mean depth of dive is between 315 and 360 m (1,024 and 1,170 ft), with a mean duration of about ten minutes. Dives can last much longer than this, at least up to two hours, with the largest whales (the adult bulls) diving deepest and longest.

Malcolm Clarke noticed that sperm whales generally reappeared at the surface at very nearly the same spot where they dived. This seemed to imply that not only did the whale probably spend a large part of the time it was submerged hanging motionless in the water, but that its descent and ascent must have been near vertical.

For a sperm whale to hang motionless in the water it would have to have neutral buoyancy, and this would vary for differing depths. If it were positively buoyant, it would rise (dead sperm whales do seem to be positively buoyant, for they float at the surface), while if it were negatively buoyant, it would sink still further. If a sperm whale were positively buoyant, it would need to swim actively to dive, using thrust from its flukes and some form of hydrofoil, probably its flippers, to drive it deeper. The only way it would have of maintaining depth would be to continue swimming, and it seems unlikely that this is what it does, in view of the observation that the whale surfaces in the same position that it dived from.

Malcolm Clarke wondered whether the spermaceti organ, for which no function had been credibly postulated by the 1970s, could act as a hydrostatic controller. Spermaceti is peculiar stuff. The ordinary fats and oils found in mammals and other animals are triglyceride esters of fatty acids, such as stearic and palmitic acids, with the trihydric alcohol glycerol. Spermaceti is a complex mixture, which, although it contains some triglyceride fats, is composed chiefly (about 73.5 per cent) of waxes. These are the esters of fatty acids and monohydric alcohols, which have a very much lower specific gravity than the fats.

Spermaceti is a clear straw-coloured oil about 30°C (86°F), but becomes cloudy if cooled below this and progressively sets and crystallises as the temperature drops. As spermaceti solidifies, it contracts and becomes denser. Malcolm Clarke found that the rate of density increase on cooling is greater at higher pressures. An increase in density of the spermaceti, of which there may be 2.5 tonnes in the head of a 30-tonne whale, will result in a decrease in the buoyancy of the creature, which, if it is already near neutral density, will cause it to sink. If, then, a sperm whale could control the temperature of its spermaceti, it could also control its density, and sink or rise with small expenditure of energy.

Malcolm Clarke set out to examine the anatomy of the sperm whale's head to see if he could discover how this control might be exerted. Anatomical investigations on whales are peculiarly difficult because of the size of the structures involved; dissection becomes more like excavation when one is dealing with a whale. Malcolm Clarke, however, managed to persuade the bone-saw man at a whaling station to cut the head of a sperm whale into manageable slices with his steam saw. From these sections, and from other studies, he was eventually able to explain how the spermaceti organ functioned.

Clarke's sections showed that the upper part of the spermaceti organ, the case, was a tough, fibrous container shaped roughly like a bullet and filled with liquid spermaceti. Beneath the case lies the junk. This is made up of a series of trapezoidal blocks of spermaceti tissue, separated by blocks of fibrous tissue. The whole series of blocks together have a form that is roughly coffin-shaped. Within and around the spermaceti organ run the nasal passages, which, when compared with those of other mammals, are extraordinarily complicated.

As in all toothed whales, there is only one external nostril, and this, as noted earlier, is situated near the extremity of the head, a little to the left side. Beneath the nostril is a shallow chamber, into which both the right and left nasal passages open. The left nasal passage runs back from this chamber, curves to pass around the left side of the case and enters the skull just in front of the brain case. The right nasal passage is quite different. To start with it runs forward to a flattened sac, the vestibular sac, at the extremity of the snout, from which it continues as a broad flattened tube which passes back through the lower part of the case to meet the left nasal passage as a narrow tube just in front of the brain case. Just before this junction it gives off another large sac, the naso-frontal sac, which lies against the crest of the skull, between it and the hind end of the case (Fig. 5.3).

Malcolm Clarke suggested that a mechanism might exist for flushing water through the right nasal passage, which, with its expanded sacs at the front and the back of the spermaceti organ, would be well fitted to act as a

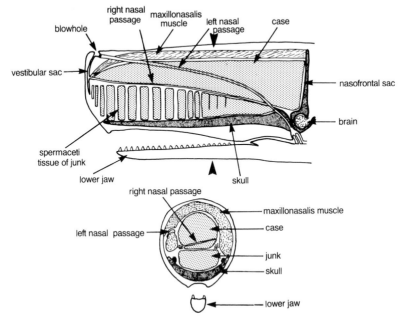

Fig. 5.3   The structure of the head of the sperm whale.

useful heat-exchanger. This would cool the spermaceti, increasing its specific gravity and causing the whale to become less buoyant.

A whale cannot 'breathe' water through its nostrils while it is submerged; there must be some way of isolating the lungs from any surface used as a heat-exchanger. This exists in the form of a sphincter muscle surrounding the right nasal passage where it joins the left, just before it enters the skull. Having isolated the right nasal passage, it can then be irrigated with water by the action of a larger block of muscle, the maxillonasalis, which can lift the forward end of the case, thus expanding the cavity within the right nasal passage. Alternatively, Clarke suggested, water might be drawn into the left nasal passage and pumped into the right, using other muscles.

The cold sea water would cool the part of the spermaceti organ in contact with the expanded lumen of the right nasal passage and its two sacs. Blood flowing through vessels in the spermaceti organ would bring more heat from within the organ, be cooled itself, and then cool oil further from the wall of the passage.

The skin of the whale is, of course, another means of losing heat. Normally the blood circulation and the blubber restrict this loss of heat, but, as explained in Chapter 1, bypasses in the blood vessels can be controlled so as to dump heat to the exterior. The blood circulation in the head of the sperm whale is designed to cool the spermaceti organ by exchanging heat from warmer incoming arterial blood and cooler outgoing venous blood.

When Malcolm Clarke calculated the rate of heat-exchange that would be possible using these two methods, he found that if the sperm whale used both together, it could adjust to neutral density in less time than it would take it to swim down to 500 m (1,625 ft). Using the skin alone, neutral density would be reached in five minutes in a dive from 200 m to 1,000 m (650–3,250 ft), while if the right nasal passage was used as well, this time would be shortened to three minutes.

The cooling of the spermaceti will cause the whale to sink. Conversely, warming it again will cause it to rise. Warming should present no problems to a submerged whale. By ceasing to circulate water through the nasal passages, and by vaso-constriction at the skin surface, it will cease to lose heat. Any activity of the trunk muscles will cause heat to accumulate and this can be transferred to the spermaceti organ via the blood. The whale will then gently rise again to the surface.

This has, I fear, been a complex (though simplified) explanation of what is a most complicated structure. It is one of the strangest stories to have emerged from the study of whales, or any mammal. Not all cetologists agree with Malcolm Clarke's theories, but they seem convincing enough to me. They are based on sound anatomical data and the many calculations that he made from his findings fit the theories and what we know of the diving behaviour of these great whales.

To use the spermaceti organ as a biological ballast tank does not necessarily mean that it may not have other uses as well. Not all its features are explained by Clarke's theories – the strange segmented nature of the junk, for example. One other suggested function is connected with what the sperm whale does once it has dived to depth.

Feeding

Sperm whales visit the abyss to exploit the food resources there. At great depths in the ocean there is no lack of life, although it is life in a very different form from the plankton and fish that live in the euphotic zone at the surface, where sufficient light penetrates to allow the plants of the plankton to synthesise organic compounds and act as the food base for all the other components of the ecosystem. Some of the deep-water forms migrate upwards to feed near the surface at night, while others are entirely dependent for food on the rain of detritus and corpses from above. Yet others are predators, like fish and squid, which feed on the smaller organisms.

Sperm whales seem to have specialised as top predators on other large deep-water predators, mainly squid but also fish. Because sperm whales have been hunted for so long, there are quite a number of records of what has been found in their stomachs. The bulk of the diet is made up of medium-sized to large squid of about a metre in length. Sperm whales also take smaller squid, down to just a few centimetres long, but they are far more famous for taking 'giant squid'. The size of giant squid, like that of sea-serpents, tends to be much exaggerated. Even the very large ones are comparatively slender-bodied and are quite small in comparison with the whale. The giant kraken of Scandinavian mythology was said to have a body one and a half miles in circumference, and arms capable of dragging the largest ship to the bottom. In a more realistic world, the largest squid reliably recorded this century was captured by a United States Coast Guard vessel on the Great Bahamas Bank (after the squid had been involved in a fight with a sperm whale). It measured 14.5 m (47 ft) overall, but this included the tentacles. Probably its body length was around 4 m (13 ft) which, while large, is small enough for a sperm whale to deal with easily. The many accounts of 'fights' between sperm whales and giant squids probably relate to the struggles of the squid which, when held in the jaws of the whale, desperately wraps its tentacles around the head of its captor as it struggles to escape. A soft-bodied animal like a squid has no chance at all against a sperm whale and it is a very unequal contest, scarcely deserving to be called a fight.

Sperm whales do not feed only on squid. In shallower waters, around island groups (they do not ordinarily approach continental shores), they also take fish. Off Iceland the large males feed extensively on lump-suckers and angler fish, while around the Azores both sexes take skate as about a quarter of their food.

Besides obvious food items, a great many other objects have been found in the stomachs of sperm whales. These include stones, sand, sponges, a glass fishing float and even coconuts and apples. It is clear that besides feeding on very active organisms like squid, sperm whales can also take objects found on the seafloor and floating at the surface.

The question now arises: how do sperm whales catch their food? The idea of an animal with the huge bulk of a sperm whale swimming actively in pursuit of such a nimble animal as a squid is not easy to accept. Not only can the squid probably outswim the whale, at least over short distances, but, being so much smaller, the squid will also be able to change direction rapidly, a thing that a whale, with its enormous mass and inertia, will not be able to do. I do not believe that the sperm whale is a pursuit hunter, at least not when it is feeding on squid at depth.

Another curious feature about the sperm whale is that it does not seem to need the whole of its jaws to feed. A considerable proportion of bull sperm whales are found with distorted jaws (Moby Dick was one of these), the jaw having been broken, presumably in fighting with another bull, and then having healed again at an angle, perhaps sticking out sideways. In such a position the jaws cannot close, but whales with such deformities are found in good nutritional condition and have obviously been feeding successfully.

Is it possible that the sperm whale uses a lure to catch its prey? It was Thomas Beale, a surgeon on a British whaler, who, in 1839, published a work on *The Natural History of the Sperm Whale*. Beale wrote:

'It appears from all I can learn among the oldest and most experienced whalers, and from the observations I have been enabled to make myself on this interesting subject, that when this whale is inclined to feed, he descends a certain depth below the surface of the ocean, and there remains in as quiet a state as possible, opening his narrow elongated mouth until the lower jaw hangs down perpendicularly, or at right angles with the body.

'The roof of his mouth, the tongue, and especially the teeth, being of a bright glistening white colour, must of course present a remarkable appearance, which seems to be the incitement by which his prey are attracted, and when a sufficient number, I am strongly led to suppose, are within the mouth, he rapidly closes his jaw and swallows the contents; which is not the only instance of animals obtaining their prey by such means, when the form of their bodies, from unwieldiness or some other cause, prevents them from securing their prey in any other manner, or by the common method of the chase.'

The twentieth century has supplied supporting evidence of the notion of the sperm whales drifting along at depth with their jaws hanging open, for, from time to time, there are reports of whales found with their lower jaws entangled in submarine telephone cables. I would add a little to Beale's suggestion. Not only is the lining of the mouth glistening white, but this contrasts strongly with the tongue, which is a bright carmine pink. I have noticed that the jiggers used by squid fishermen are often this same red-white combination, and these rely on no more than their appearance to catch squid, for they are used without bait.

But how, in the impenetrable blackness of the abyss, is such a lure seen by the squid? We do not know, but some of the deep-sea squid are bioluminescent, and may provide their own light by which to see the target. David Gaskin has even suggested that a sperm whale, having munched a luminous squid, might have sufficient light-producing slime round its mouth to make it visible.

Even if the lure serves to capture squid, which are certainly the sperm whale's principal prey, such a method is unlikely to be effective in catching the fish that have been found in sperm whales' stomachs, and it would certainly not work for inanimate objects like stones or net floats. Sperm whales must be able to pick up objects from the bottom (or surface). How this is done is not clear. The jaw is undershot (that is, the upper jaw is longer and overlaps the lower), so a direct approach seems unlikely. Perhaps the whale turns its head sideways and sucks the object into its mouth – the fish reported from sperm whale stomachs are fairly inactive species.

Another speculation about sperm-whale feeding certainly deserves mention, for it is as remarkable as Clarke's theory about the function of the spermaceti organ. It has been suggested that sperm whales stun their prey with a burst of high-intensity sound. Sperm whales can be heard to produce vocalisations of high frequency and intensity, and it was suggested that the spermaceti organ might serve as an acoustic lens to focus the sound pulses to the point at which they could immobilise a squid. This theory is discussed further in the next chapter.

The use of oxygen in deep dives

For an air-breathing mammal there are other problems concerned with diving to great depths, besides how to get down there. Humans breathe about 15 times a minute and if their breathing is interrupted for more than a minute or so, they are in acute distress. Even experienced pearl divers cannot hold their breath for more than about two and a half minutes. Clearly, mammals adapted to live in the water have to do better than this and we find that they most certainly do. Not all whales are long divers – this would serve little purpose for surface feeders, like many dolphins, which rarely dive for more than three minutes. Rorquals, however, can submerge for 40 minutes, sperm whales for 90 minutes and bottlenose whales for as long as two hours!

How whales manage to survive these long periods without breathing is a question of great interest. One thing whales do not do is take down a large store of air with them when they dive, as a human diver does. Whales do not have especially large lungs, relative to their size, although they can ventilate their lungs much more efficiently than terrestrial mammals. When a whale surfaces after a dive, it immediately exhales, discharging the lungful of air that it carried down with it when it dived, as its spout or blow. The blow of a whale has a characteristic oily smell to it, and it seems likely that it owes its visibility largely to the condensation of water on to minute oil droplets derived from the oily foam that fills the air sinus system in the head.

The blow is immediately followed by an inspiration, after which there is a pause, then another blow, and so on, until the whale has 'had his spoutings out'. During this breathing, about 90 per cent of the air in the lungs is exchanged, in contrast to some 10 or 15 per cent in a terrestrial mammal. Following a final inspiration, the whale is ready to dive again.

However, the oxygen in one lungful of air is not sufficient to sustain the whale throughout the whole of its dive. The whale has other supplies. One of the first things that one notices on seeing a whale being cut up is the darkness of the meat. The muscles of a sperm whale are very nearly black. This is due to the abundance of a protein called myoglobin in the muscles. Myoglobin, like haemoglobin, has an affinity for oxygen. The myoglobin in the muscles takes up oxygen from the haemoglobin of the blood and stores it until it is required by the cytochrome enzyme system to develop energy in the muscles.

Even this could not account for very long dives if whales had a respiratory physiology similar to our own. When carbon dioxide concentrations increase in our blood, we have an irresistible reflex response, causing us to inhale. In cetaceans there is a greatly increased tolerance to carbon dioxide levels and also to lactic acid, another respiratory metabolite. When glycogen is oxidised to carbon dioxide by the cytochrome enzymes to provide energy, the reaction can be broken at an intermediate stage when lactic acid is produced. This happens in humans when we exercise violently, and temporarily run short of oxygen, but our system will tolerate only very low levels of lactic acid. A whale, however, can store the lactic acid until it surfaces again, when it is further oxidised to carbon dioxide while the whale 'has his spoutings out'.

Besides this, it is likely that whales use the oxygen they do carry down with them in a very economical way, redistributing the blood flow to essential organs, like the brain, while less essential parts have to go into 'oxygen debt', burning their glycogen to lactic acid and no further. The blood system in whales is complicated and not very well understood. Among the most puzzling features are the *retia mirabilia*, 'wonderful networks' of contorted spirals of tiny blood vessels that form great blocks of vascular tissue on the inside walls of the thorax and elsewhere. How these function is not known. They may form a reservoir of oxygenated blood (the thoracic retia are arterial). Though they scarcely seem large enough to hold a significant amount, it should be recalled that the brain of a whale is tiny in relation to the total size of the animal.

There is another problem associated with diving, even when oxygen demands have been satisfied. This is the effect of pressure. With the increasing popularity of sport diving using an aqualung, many people are familiar with the dangers consequent upon decompression on rising to the surface too fast – that nitrogen gas, released from solution in the blood by the decreased pressure, will bubble out, causing the excruciating condition known as the 'bends', or perhaps even a fatal air embolism if the bubbles occur in a vital artery to the heart or brain.

Are whales liable to the bends? The answer is no, for a simple reason. Unlike human divers, whales never breathe air under pressure, so there is little chance for nitrogen to dissolve in the blood. A whale, as we noted earlier, dives with its lungs full of air breathed at atmospheric pressure. As it submerges, the pressure of the water around it causes its lungs to collapse; the abdominal viscera press against the oblique diaphragm, making it bulge into the thorax and take the place previously occupied by the air-filled lungs. As the lungs collapse, the walls of the alveoli, the minute chambers at the terminations of the bronchial tree, where gaseous exchange takes place, thicken and the rate of gas exchange, including the solution of nitrogen into the blood, is reduced. At a depth of 100 m (328 ft), the lungs have collapsed completely and any remaining air (or nitrogen, since the oxygen will all have been absorbed by that time) has been forced into the rigid parts of the respiratory tract, the bronchi and trachea, where little or no gas exchange can take place. Further increase of pressure will have no effect, for, apart from the minute gas spaces that remain in the respiratory and sinus systems, the whale's body is, like the water around it, virtually incompressible. There is, however, an important air space in the middle ear (see page 112), but this is protected by a blood-filled retial system bulging into it.

For the same reason that a whale does not suffer from the 'bends', it is not liable to nitrogen narcosis, either. This poisoning, by the action of nitrogen in solution on the brain, is an important limitation to air-diving by human divers, but, with just one lungful (or, more accurately, four-fifths of a lungful) of nitrogen, the whale can disregard this.

Similar mechanisms exist in those other expert diving mammals, the seals. Creatures like the elephant seals can rival the whales for depth and duration of dive. Unlike whales, however, seals empty their lungs before they dive and rely solely on oxygen carried down in solution or in combination.

Hunting

Spermaceti may be the key to the sperm whale's dominance of the deeps. It was also the cause of the relentless pursuit of the whales by an even more dangerous predator, the human being. In 1712, off the island of Nantucket in New England, Captain Christopher Hussey was cruising in a whale boat, looking for right whales to kill, when a strong north-easterly gale swept him offshore. When at last the wind calmed, he found himself in the company of a school of strange whales, with bushy, forward-pointing spouts. Hussey killed one and towed it to shore, where it was tried out. The oil from this sperm whale burnt with a bright, clear flame and none of the disagreeable odour associated with right-whale oil, while, from the solid spermaceti, fine candles could also be made.

The hunting of sperm whales soon supplanted that of right whales, which, anyway, were getting very scarce. Nantucket flourished on this trade and, by 1774, whaling vessels, mostly ships and barques, were crossing the Equator in search of sperm whales. Long voyages in deep waters made trying out the oil ashore, as had originally been the practice in Nantucket, impracticable. By the 1760s it was common practice to build the try-works on the ship, a convenient method that the Dutch and English right whalers had never adopted.

The American War of Independence checked the expansion of sperm whaling, but the trade soon picked up afterwards and, in 1789, the Pacific whaling grounds were opened up. The pioneer was a British whaler, the *Emilia*, but she was, in fact, owned by a New Englander, Samuel Enderby, who had come to England as a refugee from the newly independent United States of America. American whalers, nearly all Quaker families from New England, dominated the sperm whale fishery, British whalers tending to hunt right whales and seals. By 1842 the American fleet numbered 652 and the rest of the world had only 230 vessels. The Civil War was a blow to American whaling. Many of the whaling vessels, which came from northern ports, were captured and destroyed by the Confederate cruisers, and 40 whalers were commandeered as the famous 'Stone Fleet': filled with rocks and sunk to blockade Confederate ports.

Even these reverses were survived and Yankee whaling reached its peak around 1876, when there were 735 vessels in the trade. But decline followed. The principal cause was the depression of the market for sperm oil following the discovery of petroleum in 1859. Petroleum soon took over from whale oil as the fuel for household lamps, while coal gas lit the streets. Sperm oil was still important for lubrication in the engineering industry, however. Its quality of retaining its lubricating properties even at high temperatures and pressures, made it much in demand as a cutting oil – indeed, no real substitute has yet been found. The reduction of whales was also probably a factor in the decline of the fishery, but sperm whale stocks were never reduced in the manner of right whales.

Competition from another form of whaling, using steamers fitted with large harpoon guns in their bows (see page 169) probably finished off the sailing whalers. The last of the Yankee square-riggers to put to sea was the barque *Wanderer*. She sailed from New Bedford on 25 August 1924 and was wrecked the following day on Cuttyhunk Island. The following year two schooners, the *John R. Manta* and the *Margarett*, returned to New Bedford and Yankee whaling was part of history.

But not quite. The Yankees had exported their whaling technology far over the world, wherever their ships stopped to pick up crew or take on provisions and water. Open-boat whaling, in more or less the New England fashion, was established in places as far apart as the West Indies and Tonga. It was in the Azores, however, that the Yankee tradition was best preserved.

The seas around the Azores are good breeding grounds for sperm whales and were frequented by the Yankee whalers right up until the end of the industry. Shore-based whaling, using American methods, seems to have been established around 1832 and has continued almost up to the present day. The Azorean whalers pursued their quarry from a light wooden whale boat, or '*canoa*', about 11 m (38 ft) long and 2 m (6 ft 8 in) in beam. These were crewed by seven men, six of them pulling long oars from the thwarts, while the seventh manned a huge steering oar in the stern. In the bow was a groove for the whale line to run through and, at the stern a post, the loggerhead, around which the line could be checked. Besides oars, the boat was equipped with paddles and a mast and sail.

When whales were sighted from a look-out post on the shore, the *canoas* were launched and towed out near to the whales by a motor boat. The *canoas* then hoisted their sails and sailed down on the whales. The harpooneer, standing in the bow, hurled his harpoon into the back of one of the whales. These harpoons were made locally, but were exact copies of an American design made famous by James Temple, a New Bedford blacksmith. Once the harpoon had been hurled, the sail was lowered and the boat backed off.

Now would begin the Nantucket, or in this case Azorean, 'sleigh ride', the stricken whale towing the *canoa* behind it as it endeavoured to escape. The harpoon was connected to the whale line, some 540 m (300 fathoms) of hemp rope contained in two tubs in the bottom of the boat. The line was passed back down the boat and round the loggerhead so that it could be checked and hauled up on whenever possible so as to bring the boat within lancing range of the whale. The harpooneer was the one to lance the harpooned whale, one of the few departures from Yankee practice, which had been for the harpooneer and officer in charge of the boat, the boat-steerer, to change places at this point. If the boat could come close enough the lance would be plunged directly into the whale's body, but more usually it was hurled at the whale from a distance and then retrieved with a line.

Whales seem to have been easy enough to kill. The Azoreans took a good number of females and young, but so probably did the Yankees.

The dwarf sperm whale, *Kogia simus*. The strange 'false gill' mark is visible between the eye and the flipper.

Sperm whaling did not have the dangerous reputation of grey whaling, although, of course, accidents did happen. However, perhaps the majority of those lost in sperm whaling perished when their boats were towed out of sight by the whales.

The dead whale was towed back to the shore and stripped of its blubber and head-matter (the case and the junk), to be boiled out for oil at the whaling station. In latter years modern processing equipment was installed at some of the Azorean whaling stations and the meat and bone were also cooked out. Uniquely, the Azoreans tanned the skin of the whale's head, to make sperm whale leather.

Azorean open-boat whaling was an interesting survival. Its worst feature would seem to be the uncontrolled use of the stock without reference to what might constitute a sustainable yield. Many people would object to the killing of females and their calves, but killing males only might have had an adverse effect on the social organisation of the whales, and taking family units in a controlled manner might be the most rational way of utilising sperm whales.

It has ended now. Although there is still some demand for sperm oil, the market has disappeared since so many countries are now signatories to the Washington Convention, which limits trade in endangered species. Sperm whales are far from endangered, however, there probably being something near to two million of them, or about 80 per cent of the original stock, left in the world. But the Convention treats all whales as one and bans trade in their products.

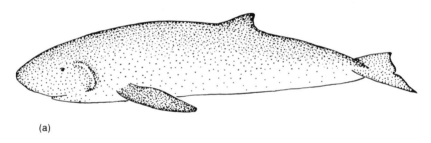

(a)

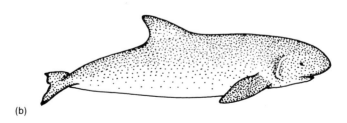

(b)

Fig. 5.4   Relations of the sperm whale: (a) the pygmy sperm whales, *Kogia breviceps*; (b) the dwarf sperm whale, *K. simus*.

The genus *Kogia*

Before leaving the subject of sperm whales, mention must be made of two small and rather insignificant cousins of Moby Dick. The first of these, the pygmy sperm whale, *Kogia breviceps* (Fig. 5.4a), is quite unlike its larger relative. Reaching a length of about 3.4 m (11 ft), the males and females are about the same size. They look vaguely shark-like, for there is a strange sort of false gill on the side of the head, which is proportionately much smaller than that of the sperm whale, though having something of the same structure. Pygmy sperm whales are very rarely observed at sea, although they strand fairly frequently. They appear to be cosmopolitan, occurring in both temperate and tropical waters, although not in polar regions.

Even smaller than the pygmy sperm whale, and often confused with it, is the dwarf sperm whale, *Kogia simus* (Fig. 5.4b). This reaches a maximum length of 2.7 m (8 ft 9 in). There are seven to twelve pairs of teeth in the lower jaw (12–16 pairs in the pygmy) and, as in the sperm whale, no functional teeth in the upper jaw.

Both these small whales feed principally on squid, but crustaceans and fish are also taken. Stomach contents indicate that they can dive to at least 300 m (975 ft). The pygmy sperm whale is the second most commonly stranded cetacean (after the bottlenose dolphin) in the south-eastern United States, but we know very little about these two small whales.

Chapter 6
The Oceanic Dolphins

For most people the best chance to see a living cetacean at close quarters in its natural environment, will come as an encounter with a dolphin of one sort or another. The sight of a school of dolphins at sea is always exhilarating. This is what Herman Melville had to say:

> 'Their appearance is generally hailed with delight by the mariner. Full of fine spirits, they invariably come from the breezy billows to windward. They are the lads that always live before the wind. They are accounted a lucky omen. If you yourself can withstand three cheers at beholding these vivacious fish, then heaven help ye; the spirit of godly gamesomeness is not in ye.'

Often dolphins approach close to ships, particularly sailing boats, and one or two of them may take up station in the bow wave, hanging apparently motionless, but keeping up effortlessly with the vessel's movement. Just how dolphins do this has always excited the curiosity of sailors, but it has only been in recent years that we have been able to explain how the dolphin hitches its ride.

How dolphins swim

When a ship moves through the water it creates a field of positive pressure just ahead of, and to the side of, its bows. The bow wave that runs away from the stem of the vessel is the visible demonstration of the release of this pressure field at the air-water interface. If the dolphin positions itself so that its flukes are against the pressure field (Fig. 6.1), it will be propelled forwards because of the difference in pressure on either side of its flukes – a situation not very different from that of a surfboarder riding a wave. It is not necessary for the dolphin's flukes to be at the surface for this to take place. Provided there is the requisite pressure difference, then the animal can take advantage of it. And bow-wave riding need not be restricted to ships, either. Any object moving through the water creates a pressure field; small dolphins have been observed getting a free ride in the pressure field created by a larger companion, and it may be common for young dolphins to travel alongside their mothers like this, thus maintaining close contact and saving some of the energy of locomotion. This may even be the origin of the habit of riding in ships' bow waves – such a wave mimics the pressure field set up by a mother earlier in life and to the dolphin this perhaps contributes a sense of security.

But dolphins are not usually passive riders. They are active swimmers capable of astonishing feats of speed and agility. Rather few people may have been privileged to have seen a dolphin at large in the limitless ocean, but many of us have visited a dolphinarium and seen the animals on show

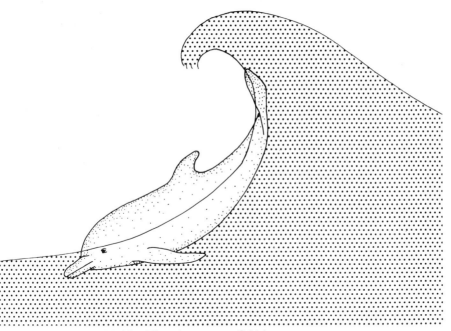

Fig. 6.1　A dolphin riding the pressure field in a wave crest.

swimming vigorously around their pool and, at the whistle of their trainer, leaping out of the water, pirouetting as they do so, to catch a hoop on their beaks, or some other such trivial and rather demeaning antic.

A well-run dolphinarium can provide excellent opportunities for studying the actions of swimming in these small whales, and the process is not greatly different from one end of the size spectrum of the Cetacea to the other. Watching a dolphin move through the water, it is easy to see that the effective organ of propulsion is the flukes and the hind third of the body, which are moved up and down in a vertical plane, developing the thrust that drives the animal along (Fig. 6.2). The flukes have evolved in

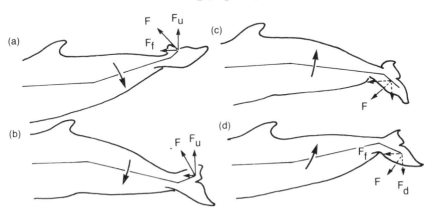

Fig. 6.2　The forces acting on the flukes of a dolphin. As the tail beats up and down a forward thrust is developed that drives the dolphin forward.
(F = resultant force, F_f = forward force, F_u = upward force, F_d = downward force.)

cetaceans specifically to provide a surface to push against the water. They are anatomically fibrous lateral expansions of the extreme end of the tail with a rigid, but quite elastic, structure. In order to produce the required effect, as we shall see later, the flukes have to alter their angle of movement against the water at different phases of the beat of the tail. This requirement is provided for automatically by the structure of the flukes themselves, and by the way they are attached to the rest of the tail.

The tough and dense fibrous core of the flukes, which is what attaches them to the vertebrae of the tail, is surrounded by an envelope of bundles of ligaments arranged so as to resist bending either up or down. If one looks at a section of a fluke (Fig. 6.3), it is possible to see that these fibres are pleated, with those on the lower surface having more pleats than those

Fig. 6.3 Sections through the flukes of a dolphin: (a) at rest; (b) at upward power stroke; (c) at recovery downstroke.

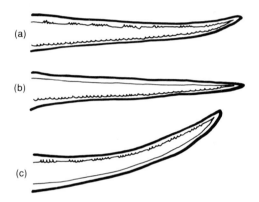

on the upper side. This means that if the fluke is raised upwards against the resistance of the water, the rather sparse pleats on the upper side will allow only a minimal change of curvature of the fluke, while if it is moved downwards, the more numerous pleats on the lower side will open up, allowing considerable bending of the fluke to occur.

Moving the flukes up and, to a lesser extent, down, is what actually drives the dolphin through the water. In fact, it is the whole tail that is moved, and looking at the external shape of a dolphin, or any whale, it is easy not to realise what actually constitutes the tail. In vertebrates, the tail is the whole region of the body posterior to the anus. Usually this is quite easy to recognise in mammals, being a mere appendage to the trunk. But in whales it is different. Here the anus is about two-thirds of the way back, and a substantial part of the body lies posterior to it. Yet this is truly the tail, and it is the flukes that are the appendage attached to it. It is the concentration of the main muscle masses used in locomotion into blocks along the spinal column and their extension into the anatomical tail that has led to this fish-like and rather unmammalian appearance.

The body of a dolphin is not very flexible (and that of the larger whales even less so). What mobility there is is mainly concentrated at the base of the tail (Fig. 6.4), with a much more restricted range of movement at the neck. The beating of the tail is brought about by two pairs of muscle

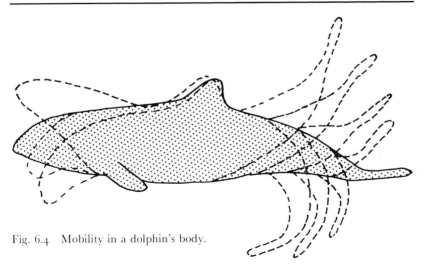

Fig. 6.4 Mobility in a dolphin's body.

blocks, running above and below the transverse processes of the vertebrae. Of these, those above, the epaxial muscles, are much larger than the hypaxial muscles below (Fig. 6.5). This difference in size hints to us that it is the upward movement of the tail, brought about by the epaxial muscles, that is the driving force, or power stroke, in swimming.

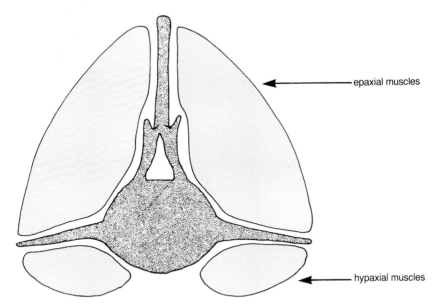

epaxial muscles

hypaxial muscles

Fig. 6.5 Muscle blocks around a dolphin's vertebra. The epaxial muscles, above the vertebral transverse process, which are responsible for the power up-stroke of the tail, are much bigger than the hypaxial muscles below, which pull the tail down in the recovery stroke.

The speed that whales can reach is impressive. Large rorquals cruise at about 9.25 km/hr (5 knots) or less, but, when alarmed, can swim at 26–27.7 km/hr (14–15 knots) or even sprint at 37 km/hr (20 knots) for short distances. The sei whale, which is probably the speediest of all the rorquals, has been reported to reach as much as 64.7 km/hr (35 knots) for short distances. Generally speaking, for propulsion in water, larger bodies using the same power system will travel faster than smaller ones, and we would expect to find that dolphins would be much less speedy swimmers than the great whales. But this is not the case. The speeds that dolphins can reach can be rather easily calculated by watching them leap out of the water, as they often do, for the exit speed to reach a certain height of leap can be calculated quite simply. A dolphin jumping about 5.5 m (18 ft 7 in) out of the water will need to be travelling at 36 km/hr (20 knots) or 10.5 metres per second, just before it leaves the water, and this is quite comparable with the speed of the much larger rorquals.

This observation has puzzled biologists and ship designers for many years. In ship design the waterline length dictates the maximum speed or the vessel. Install additional power for the engines above a certain maximum and there will be no increase in speed – the extra power will be wasted as turbulence around the propellor or hull. It is this matter of turbulence that is the critical factor when considering the movement of bodies, be they ships or dolphins, through the water. When a body is moved through water, the water molecules nearest the body surface are drawn along with it, creating *drag*. Layers of water a little further from the surface will move more slowly until, at a certain distance, the water is undisturbed. In ideal conditions, if the body is suitably streamlined, the layers will flow smoothly over each other in a state of *laminar flow*. If the body is not correctly shaped, or if the speed of movement through the water increases, the laminar flow breaks down into a series of eddies, *turbulence*, and much energy is absorbed. The nearer the front of the body the eddies develop, the greater the retarding effect. The function of streamlining is to create flow lines that carry the turbulence away to the rear where it has little effect.

What the dolphin has done is to eliminate turbulence. Laminar flow exists over the entire body of a swimming dolphin and the power requirement for high speed swimming is radically reduced. A stationary dolphin, floating nearly horizontal in the water, begins to move forward. The power stroke is the upward movement of the tail, brought about by the epaxial muscles. As the flukes come up, water is forced from their upper to their lower surface; this water flow is unable to pass easily over the sharp trailing edge of the flukes, turbulence is set up and a vortex develops (Fig. 6.6a). Further upward movement of the flukes creates an area of low pressure beneath them, and water is drawn from the surface of the head and chest. As a result the dolphin moves forwards and downwards, against the hydroplaning action of the flippers (Fig. 6.6b). Once forward and downward motion is established, further upward travel of the flukes accelerates water obliquely over the body, past the fared edge of the back behind the dorsal fin, washing away the vortex turbulence at the trailing edge of the flukes (Fig. 6.6c).

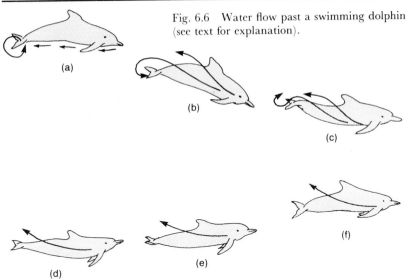

Fig. 6.6 Water flow past a swimming dolphin (see text for explanation).

(a)

(b)

(c)

(d)

(e)

(f)

The dolphin has now reached its maximum velocity for this power stroke and the flukes are at their highest. They adopt a neutral glide position (because of their elasticity) and offer negligible resistance to the water (Fig. 6.6d). The hypaxial muscles now begin to pull the tail downwards, ready for the next power stroke. The tips of the flukes curl upwards, spilling water sideways instead of to the rear. The buoyancy of the oil-rich head and chest region causes the front of the body to rise as the tail descends at about half the rate of the power stroke (Fig. 6.6e). At the end of the recovery stroke, the position of the dolphin and the attitude of its tail are approximately the same as at the beginning of the power stroke, and the cycle is repeated.

In the swimming dolphin, turbulence is actively avoided by washing away vortices as they are formed by the swimming movements. Streamlining is, of course, vital in this, but a dolphin is not rigidly streamlined, like a submarine. Its body shape can adapt to water flow so as to avoid turbulence; in other words, it is dynamically streamlined. When a dolphin is swimming fast, transverse corrugations appear at right angles to the direction of movement through the water. These mark areas where incipient turbulence is developing, but, instead of going on to produces vortices, as would be the case in a rigid-hulled ship, the shape of the dolphin responds to the pressure differences, eliminating the turbulence as it is formed. Dolphins may be even cleverer than this. It has been suggested that a swimming dolphin continuously sheds skin cells, together with oil-droplets, from the surface of its body, and these may serve to reduce drag, as does the slime secreted by fish.

Intelligence in dolphins

Most sightings of dolphins at sea are unlikely to impress the observer with the intelligence of these animals. However beautifully the dolphin may appear matched to its environment, however elegantly it floats in the bow wave, the range of the behaviour that it displays can tell us little of its mental processes. Things are very different in captivity. The range of tricks that dolphins can be taught, the speed with which they acquire their skills and their evident capacity to invent new displays, all suggest a creature with considerable mental powers.

Is this suggestion justified? Are dolphins, or other cetaceans, endowed with an intelligence second only to that of humans? These questions are intrinsically difficult to answer. No aspect of an animal's biology is more difficult to examine than its mental processes. Furthermore, we have difficulty enough in measuring intelligence in our own species, without the great extension required if we are to make meaningful comparisons with other species, particularly those which evolution has equipped for life in a vastly different medium.

A crude measure of intelligence might relate to brain size. Whales and their kin certainly have very large brains. The brain of the sperm whale, at 7.8 kg (17 lb 3 oz) is the largest brain known. Fin whales are not much less at 6.93 kg (15 lb 4 oz), but we have few observations of what might be termed intelligent actions in either of these. The common dolphin has a brain weighing about 0.88 kg (1 lb 15 oz), which is not really much less than the human brain of 1.3–1.7 kg (2 lb 14 oz–3 lb 12 oz).

But brain size, of course, is related to absolute body size, although not directly. Is there some other anatomical feature that might tell us more about the creature's intelligence? Such a feature might be the relative size and degree of folding of the cerebral hemispheres, where we know from our own species that many of the higher processes go on. Indeed, we find that the cetacean brain does have very large cerebral hemispheres and they are highly convoluted. It would seem possible that cetaceans have large areas of brain available for higher mental functions.

What happens if we look at the behaviour of whales and dolphins? There is no doubt that dolphins can be trained to carry out remarkably complicated routines. But in the hands of skilled trainers this is true also of troupes of dogs, or even pigeons, in a circus. It has been suggested that dolphins have a language by which they can communicate with each other. An experiment was carried out where a pair of bottlenose dolphins were placed in a pool divided by a barrier through which they could not see, but could hear, each other. The first dolphin could see a pair of lights, one of which would be flashed, while the other had to press one of a pair of plates, as indicated by the appropriate, unseen, light, in order to obtain a food reward for both animals. The dolphins soon learned to do this, and it was deduced that they had a communicative language. This would certainly be evidence of intelligence of a high order. However, the experiment seemed less remarkable when, as Harrison Matthews pointed out, a pair of pigeons (not a species renowned for its intelligence) was trained to perform the same trick. We should never underestimate the skill of trainers in examples such as this, and rigorous experimental procedures are needed to prevent trainers 'cueing' their subjects.

Dolphins certainly have a highly developed imitative sense. When a young bottlenose dolphin was placed in a pool containing a spinner dolphin, it immediately made a very creditable attempt to imitate the spinner's highly characteristic spin when it leapt out of the water. The young bottlenose had not seen a spinner before, but, without prompting, it copied its action. We may recall that imitativeness is a characteristic of our own species and several of the higher apes. What role it might serve in an aquatic mammal like a dolphin is not obvious.

Dolphins are sociable animals and, with other toothed whales, show what appears to be concern for injured members of their group, for example, supporting them at the surface. Dolphins in captivity have been observed to act as midwives when one of their group gave birth, assisting the newborn calf to the surface. Many people suppose that this care-giving behaviour indicates a sense of compassion in the whales and that this must be associated with high intelligence. This may not be the case. There is some evidence that this type of behaviour can be very stereotyped and unrelated to any feeling as complicated as compassion, as when a dolphin in captivity spent many hours supporting at the surface a small shark which it had itself killed shortly before.

The many tales, from classical times to the present day, of dolphins saving people lost at sea by supporting them at the surface, may relate to this behaviour. Some dolphins seem actively to seek out human company. Such sociable dolphins seem often to be animals that are not associated with a normal group structure and concentrate on certain stretches of inshore waters. Such an animal was Beaky, a bottlenose dolphin that frequented the Isle of Man and the south-west coast of England between 1972 and 1978. Beaky, like many of these sociable dolphins, was a male, and he seemed to select for his playmates women and children. To women he was overtly sexual, rubbing his body against them and hooking his erect penis round their waists. He seemed particularly to relish reciprocal caresses from women and would linger with them if they in turn stroked his sides. To a man Beaky would make the same advances, but should the man stroke back, Beaky's behaviour changed abruptly, seeming to demonstrate that he was exhibiting male dominance, rather than simply sexual feelings. To children, Beaky was reported to exhibit no sexual behaviour.

What interests me particularly in this is not so much the transference of the dolphin's social behaviour from its own species to humans, but the fact that it could evidently distinguish between adult males, females and children. This to me seems a remarkable and unexplained faculty.

Some people believe that dolphins have a complicated language and that it might even be possible for people to communicate with them, perhaps by using a sign language that the dolphins could be taught and with which they could express their own tongue. Experiments to test this in dolphins (and in apes) have yielded no clear-cut results and we do not seem much nearer to communication with either dolphins or chimpanzees.

A feature of cetacean behaviour that does not indicate to me the possession of a high level of intelligence is the puzzling phenomenon of mass strandings. From time to time groups or whole shoals of dolphins are

Beaky, a friendly bottlenose dolphin, swims past a spectator in Cornwall.

stranded on the shore. Many strandings are the result of accident, as when an oceanic species finds itself in shoal waters, or of sickness or death, where the animal drifts ashore passively. These are explicable enough. But there are occasions when, for example, a group of pilot whales, in an apparently healthy condition, simply swim ashore. If people attempt to pull them out to sea, they swim back to strand again with their companions.

What can we make of this? The answer will ultimately depend on what is meant by intelligence and how far we are prepared to bend any definition of this quality used for our own species, or for apes, so as to fit a very differently organised mammal living in water. I do not consider that cetacea have exceptional mental powers. They are capable of some fine discrimination, are imitative and (hence) easily trained. But they are also captivating animals and thus people tend to attribute to them qualitities they do not possess.

Sight

We might then ask what it is that dolphins actually do with their large and complex folded brains. One thing they do is use their brains to receive and process information about their surroundings, and, from this, to modify their behaviour. Because dolphins live in water, they have to rely on rather different sensory systems from those we are familiar with in ourselves or other terrestrial mammals. Humans, and other primates, are very dependent on a sense of sight. We are diurnal, and there is usually plenty of light about to provide us with information about our environment. In water, conditions are different. Even in clear sea water, 90 per cent of the light has been absorbed at a depth of 10 m (32 ft 6 in), and at 40 m (130 ft) 99 per cent has gone. By 400 m (1,300 ft) it is pitch black, apart from the pale bioluminescence of some of the creatures of the deep. The maximum range of visibility in the sea, even under good conditions, is likely to be no more than 60 m (195 ft), so a big whale might not even be able to see its own flukes!

But whales (except for some freshwater dolphins) do have functional eyes, although they have been much modified to resist the pressure and sheer as the animal travels under the water, and the refractive system has been modified so that a sharp image can be formed when submerged. A dolphin beneath the surface forms a sharp image of its surroundings from the refraction of its nearly spherical lens alone, but when its head emerges, the extra refraction at the interface between the cornea and the air causes it to be very short-sighted. It is the converse of the human diver seeing clearly in air, but being unable to form a sharp image when submerged unless wearing a face-mask. Dolphins and large whales certainly use their eyes to look around the surface, an activity known as 'spy-hopping'. Probably they correct their short-sightedness by contracting the pupil of their eye, thereby reducing the aperture of the lens, which, as every photographer knows, increases the depth of focus. The different intensities of light below and above the water will act in the favour of the whale in this response, just as it acts against the human diver.

Hearing

A much more important source of information to a cetacean comes from the sounds it hears. Sound travels faster in water than in air and, more important, it is attenuated far less, so that sounds are audible (to a properly adapted ear) over much greater distances under water than in air. Human divers, putting their heads under the water, may not think that the sounds they hear are especially informative. This is because directional hearing is lost when a terrestrial mammal submerges its head. We can locate sounds because our ears open on the sides of our heads – some animals may increase the effective separation and directionality of their hearing by having large, movable, external ears. Comparisons within the brain of the sounds received by each ear allow some localisation of their origin. Under water, however, sound enters the ear not only through the auditory canal but also through the tissues of the head, and particularly the bone of the skull, passing directly to the inner ear.

Cetacean ears have been modified so as retain their power of localising sound under water. The hearing apparatus, instead of being embedded within the bones of the skull, is joined to it only by a slender process. The bones that surround the inner and middle ear, the petrosal and tympanic bulla, are themselves surrounded by sinuses filled with a fine foam of gas bubbles in an oil-mucus emulsion. This is a good sound insulator and so, although sound waves enter the whole head of the whale, just as they do the human diver, the inner ear remains isolated, except for the narrow connection of the auditory meatus. This is not an open tube, as in terrestrial mammals, but rather a thread of tissue (in mysticetes) or a narrow tube (in toothed whales). Despite its small size, it has been shown, by means of acoustic probes, that this is indeed the route through which sound reaches the whale's ear.

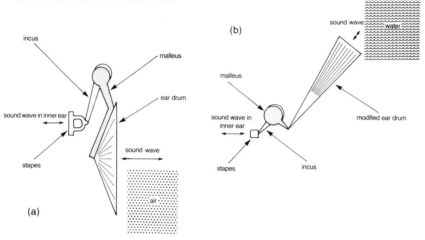

Fig. 6.7 The lever system of the bones in the middle ear in : (a) a terrestrial mammal and (b) a dolphin. The difference in the lengths of the levers ensures that the displacement amplitude of sound waves in air and water is compensated for, so that the inner ear receives the same wave amplitude and pressure.

The middle ear in whales has also been much modified. The system of bones or ossicles that is found in all mammals – the malleus attached to the ear drum, the incus that connects the malleus to the stapes, and the stapes itself, which transmits the sound vibration to the oval window of the cochlea – are present in cetacea also (Fig. 6.7). However, because the displacement amplitude of sound vibrations in water is about 60 times smaller than that of sound transmitted in air, and the pressure amplitude is about 60 times greater, although the cetacean inner ear is virtually the same as in other mammals, some adjustments have to be made. This is done by altering the lengths of the lever system made up from the three ear ossicles, the malleus, incus and stapes, so that the vibration arriving at the oval window, although originating in water, has the same character as vibrations arising in air.

The complexity and perfection of the cetacean auditory system is a thing to marvel at. Much of what we know of it has been derived from the work of two British cetologists, Francis Fraser and Peter Purves. Besides what I have summarised above, they showed also how the arrangement of the bones prevented resonant vibrations, arising in the skull, being passed to the inner ear; how the auditory meatus could be moved in and out of acoustic shadows to facilitate binaural hearing; and even how the whole system is protected from damage, at depth, by blood-filled tissue that could expand into gas spaces in the middle ear when these were compressed as the whale dived.

Vocalisation and echo-location

But dolphins, of course, are not merely passive hearers. This is what Aristotle had to say in the fourth century BC:

'The dolphin, when taken out of the water, gives a squeak and moans in air.... For this creature has a voice, for it is furnished with a lung and a windpipe; but its tongue is not loose, nor has it lips, so as to give utterance to an articulate sound.'

Among the many sounds produced by cetaceans, the squeaks, or creaks, are of particular interest. If these sounds are analysed, it is found that they are composed of a series of very brief pulses of sound, or clicks, each lasting for perhaps only a fraction of a millisecond. These clicks are produced in trains of around 40–50 pulses per second, though this can vary greatly. It is when the pulse rate rises to several hundred a second that we can perceive the sound as the creak of a door hinge.

These sounds can convey a great deal of information. Not only are they different in different species, but within a species different individuals may produce characteristic vocalisations. John Dreher, in the United States, studied six typical whistles of the bottlenose dolphin by playing back recordings to the dolphins. The first vocalistaion, which he called an 'upward glide', represented a 'search' vocalisation and was a common response to any new stimulus. A 'downward glide' was most often heard in a distress situation. A rising-falling-rising whistle caused the dolphins to become highly excited and one male erected his penis. Rise-fall-rise-fall seemed to be associated with excitement and irritation, while a fall-rise-fall subdued the activity of the dolphins in the pool.

On should be wary of reading too much into these results. These whistles are not speech, in the sense of a learnt medium for communication. They are present in the bottlenose dolphin on the day of birth, albeit in a slightly ragged state. Perhaps the most remarkable use of the vocalisations of dolphins is in orientation by echo-location, or sonar.

Echo-location depends on sending out a pulse of sound which is reflected from a target and the time taken for the echo to return is then measured. The velocity of sound in water is constant, so that the distance of the target can be calculated. An echo-sounder on a ship makes these measurements electronically; a dolphin does the calculations in its brain. To a mere human, the precision that a dolphin, or a bat (another group of mammals that have refined echo-location powers), can achieve by echo-location is amazing. Not only can dolphins avoid obstacles in the dark, or locate food entirely by echo-location, but they can discriminate between fish of different sizes. A blindfolded dolphin was shown to be capable of discriminating between metal spheres that differed in diameter by only 9.5 mm ($\frac{3}{8}$ in), or 15 per cent! When making fine discriminations like this, the dolphin emits a longer and more varied series of clicks – it is interrogating its environment!

For echo-location to be effective the sound energy has to be directed to the target. Ken Norris studied dolphins in California and from his observations he suggested that the clicks were produced not in the larynx, but just below the blowhole, in the region of the nasal plugs (Fig. 6.8). The sound would then be reflected from the bony surface of the skull, which would act as a parabolic reflector, producing a focused beam. As a refinement, Norris suggested that the melon might act as an acoustic lens, concentrating the sound energy in a narrow beam. Norris's theory goes further and suggests that the lower jaw acts as an acoustic probe, picking up the returning echoes and conducting them directly to the inner ear, via the jaw articulation

This is an ingenious theory, but it is not accepted by all scientists. Peter Purves considers that the vocalisations are produced from the larynx, in the conventional mammalian manner, that the front of the skull (which is not parabolic, nor even concave in some whales) does not act as a reflector, and that the melon has no acoustic role. It is possible that sound energy is transmitted to the water by the bony structure of the skull and its associated cartilage. The matter remains unsettled, but I, at any rate, am convinced by Purves's arguments.

It is perhaps this additional echo-locating sense that is responsible for the very large brain size of cetaceans. In the nerve connecting the ear with

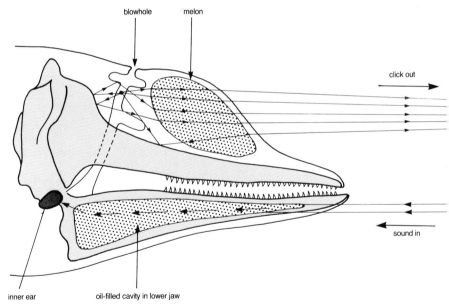

Fig. 6.8 Sound propagation and reception in a dolphin, according to Ken Norris. The melon acts as an acoustic lens, focusing the clicks, while the oil-filled cavity of the lower jaw acts as a wave guide to conduct the sound received to the ear.

the brain in the bottlenose dolphin, there are about 67,900 fibres, more than twice as many as in a human, and substantially more than in a cat, an animal with a very acute sense of hearing. Similarly, those parts of the brain involved in sound perception are greatly developed. Part of the function of the complexly folded cortex of the dolphin's brain may be to handle the very large load of information which it receives from its sense organs, in particular its ears. Acoustic information is believed to require a lot of brain tissue for processing.

Ken Norris, working with a Danish colleague, Bertil Møhl, suggests that another faculty evolved alongside echo-location. This is the ability to use a range-and-direction-finding sound beam, greatly amplified, to stun prey. It has long been a puzzle how some of the odontocetes (not particularly the dolphins, which mostly have pretty useful teeth, but some of the less well-endowed species, like the beaked whales) actually catch their prey. A sound beam can, in theory, carry plenty of energy – researchers have killed anchovies with sounds similar to those recorded in the vicinity of whales – and could be used at a distance, because of the small attenuation of sound in water. The concept of an underwater death beam is intriguing (and California seems appropriate for so bizarre a suggestion) but so far there is little hard evidence to support this theory. Such sound pulses should be both frequent and easily detected if dolphins, sperm whales and beaked whales are using them routinely to catch their food, yet none have yet been unequivocally recorded in the wild from the vicinity of feeding whales (or elsewhere), nor in the closely studied captive dolphins. Research is currently going on at the Long Marine Laboratory in Santa Cruz to test this theory which, inevitably, has been dubbed the 'big bang'. I shall reserve judgement until, instead of speculation, some firm data are available.

Stranding

Another hypothesis for which there is better support, though still no conclusive proof, concerns whale navigation and the question of strandings. Strandings of whales have provoked wonderment and curiosity for centuries. In the United Kingdom, where whales are 'royal fish' and the property of the Crown, there is an excellent record of strandings around the coasts. Whales which die at sea, from natural causes or accident, may bloat and drift about until they fetch up on a shore, which may be anywhere. Live strandings, which are much less frequent – only 137 out of around 3,000 records in the British series – show a distribution that is related to an important, but little known, geophysical pattern – the total geomagnetic field of the Earth. Margaret Klinowska has shown that all the British live strandings occur where geomagnetic contours cross the coastline at right angles. She suggests that the whales use the total geomagnetic field as a map, using not the directional differences (as we do with a compass), but the small relative differences in total field.

The total geomagnetic field fluctuates in a fairly regular manner each day, and it is possible that whales use these fluctuations as a timer to tell how long they have been travelling, coupled with a learnt or instinctive tendency to follow a field of constant strength, that is, to follow a geomagnetic contour. This ability might be the basis of a system of whale navigation.

Klinowska noticed that live strandings tended to occur on days on which the daily fluctuation had been obscured by solar activity or other irregular changes. On the south coast of Britain, strandings occurred two days after a magnetically disturbed morning, while on the north coast it was about a day and a half later. Looking at the geomagnetic map, she discovered that there were two major geomagnetic crossroads near the United Kingdom, one about a day and a half's swim from the Scottish coast, the other about two days from the south coast. This suggested to Klinowska that the doomed whales made their mistake while still some distance from land, but then swam on regardless, following a contour till it lead them, not past the coast, but straight onshore. This theory could account for the often observed fact that beached whales, when towed back out to sea, swim to the shore once more. They are, it seems, convinced that is the right direction in which to travel.

If whales do use the geomagnetic field to navigate, they would need a detector. This might be magnetite, a magnetic oxide of iron, which has been shown to occur in the head of the common dolphin (as well as in the pigeon, which may also navigate magnetically). However, magnetite would be more likely to give information on field direction, whereas Margaret Klinowska's theory demands a determination of relative field strength. Like the 'big bang', this theory is also the subject of current research.

So far in this chapter I have discussed some of the common characterishes of dolphins (and other whales). Now it is time to look a little more closely at the various species that comprise this, the largest of the groups of the toothed whales. The superfamily that includes the dolphins is divided into three families, two rather similar and one very different. The very different one, the Monodontidae (the narwhal and beluga) I shall deal with in a separate chapter (Chapter 8), and in the dolphin family I shall take the subfamily that includes the killer whale and its kin and deal with that separately too (Chapter 7). This leaves the family of the porpoises, and four dolphin subfamilies, the rough-toothed dolphins, the true dolphins, the right-whale dolphins and a group of four small dolphins with no common collective name.

The porpoises

The porpoises (Fig. 6.9) are characterised by their spade-shaped teeth and absence of a beak. 'Porpoise' tends to be used in a general sense in North America to mean any small dolphin, but it is a useful term to separate the six rather similar species in this family. The harbour porpoise, *Phocoena phocoena*, is a familiar little cetacean in coastal waters of the North Atlantic. Unfortunately, it is a good deal less common than it once was and no one really knows the reason for this, whether it be pollution, dis-

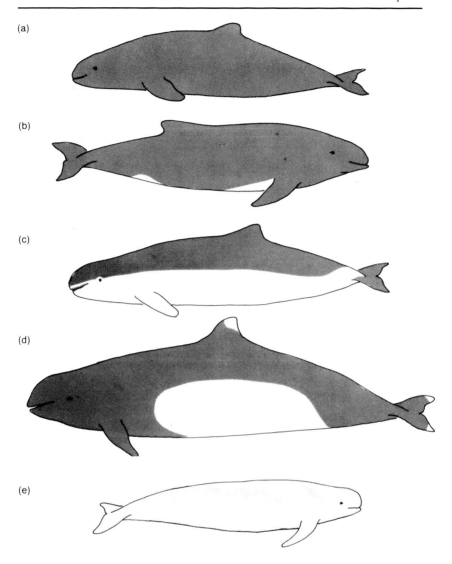

Fig. 6.9 Porpoises: (a) harbour porpoise, *Phocoena phocoena*; (b) Burmeister's, *P. spinipinnis*; (c) spectacled, *Australophocaena dioptrica*; (d) Dall's, *Phocoenoides dalli*; and (e) finless, *Neophocaena phocaenoides*.

turbance or conflict with fisheries. It reaches a length of only about 1.8 m (6 ft), males being slightly smaller. As in most members of this family, the head slopes up to a low, receding forehead. There are 22–27 pairs of spade-shaped teeth in both jaws. A large triangular dorsal fin is situated about the middle of the back. It is usually dark grey above and paler beneath.

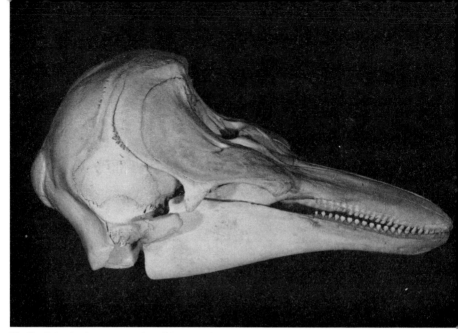

Skull of the harbour porpoise, *Phocoena phocoena*. The spade-shaped teeth contrast with the needle-like teeth of the true dolphins.

Although a familiar sight, remarkably little is known about the harbour porpoise. It seems to enjoy a varied diet, taking some 3–5 kg (6.6–11 lb) per day of whatever fish are generally available. They become sexually mature at five to six years (a year earlier in the west Atlantic), and seem not to form lasting pairs. Lactation probably continues for six to eight months and the animals live for 12–13 years. Porpoises are not particularly gregarious and groups of two to four animals are usual, as are solitary specimens.

The cochito, *P. sinus*, from the Gulf of California, is a very poorly known smaller local version of the harbour porpoise. Burmeister's porpoise, *P. spinipinnis*, from the Atlantic and Pacific coasts of South America, is another similar form. It has fewer teeth, some 14–16 pairs in the upper, and 17–19 pairs in the lower jaw. It is caught in great numbers in gill nets off Peru and Uruguay and is considered to be, like the cochito, a vulnerable species. The spectacled porpoise, *Australophocaena dioptrica*, from the western South Atlantic, has a more distinct colour pattern, black above and white below, with the white encircling the eyes to give the spectacled appearance. It appears to be uncommon.

Dall's porpoise, *Phocoenoides dalli*, is an oceanic porpoise of the waters over the continental slope of the North Pacific. It is larger than the porpoises already described, about 2.5 m (8 ft) long, black, with a conspicuous white patch on its flanks and belly, and white tips to its fin and flukes, although these vary sufficiently for four different forms to be recognised. Its population has been estimated as between 790,000 and 1,738,000, being particularly common in the southern Bering Sea and the Sea of Okhotsk. However, several thousand are taken yearly for food in Japan and more than ten thousand are drowned accidentally each year in Japanese gill nets and discarded.

The remaining porpoise is the odd one in the group. This is the finless porpoise, *Neophocaena phocaenoides*, which is found along the coasts and rivers of Pakistan, India and the whole of South-east Asia north to China, Japan and Korea. As the name suggests, there is no dorsal fin, although the back is ridged. It is a widely distributed porpoise, but does not appear to be particularly common anywhere.

Rough-toothed dolphins

The four species comprising the rough-toothed dolphins (Fig. 6.10) are grouped together because of specialities in the air sinus system in their skulls. The rough-toothed dolphin, *Steno bredanensis*, is a medium-sized dolphin, up to 2.75 m (9 ft), found in tropical and subtropical waters of the Atlantic and Indo-Pacific. The 20–27 pairs of teeth are peculiar in being furrowed on their crowns, although this is not conspicuous. The rough-toothed dolphin is dark grey on the back and flanks, with a white throat and belly and often blotches on its sides.

The tucuxi, *Sotalia fluviatilis*, is found in the large rivers and along the Atlantic coast of South America. It is the smallest of all the cetacea, reaching about 1.8 m (6 ft) in length, although it can mature at 146 cm (4 ft 9 in). The tucuxi is an abundant species in most parts of its range, but like the true river dolphins (Chapter 9), it faces threats from human encroachment.

There are two species of humpbacked dolphins, one from the Indo-Pacific region, *Sousa chinensis*, and one from the Atlantic, *S. teuszii*. Both are

The Indo-Pacific humpback dolphin, *Sousa chinensis*.

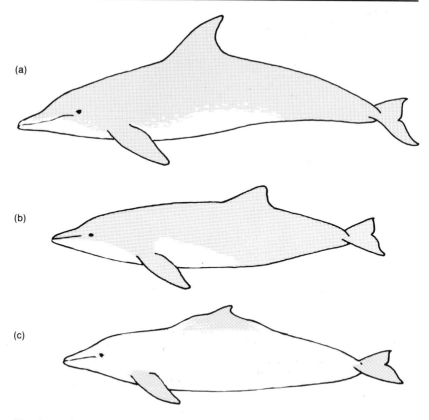

Fig. 6.10   Rough-toothed dolphins: (a) rough-toothed dolphin, *Steno bredanensis*; (b) tucuxi, *Sotalia fluviatilis*; (c) Indo-Pacific humpbacked dolphin, *Sousa chinensis*.

warm-water dolphins, characterised by the striking hump on which their dorsal fins are borne. Like the other two species in this group, they have a well-developed beak. The Indo-Pacific humpbacked dolphin may become very pale, almost white, with age, but some populations may be darker or even spotted. They grow to about 2.8 m (9 ft). It has often been stated that the humpbacked dolphin feeds on vegetation, but this is undoubtedly due to confusion with manatees.

The common dolphin

The true dolphins comprise about 14 species. We cannot be sure about the actual number of distinct forms because there is still disagreement between various experts on how some populations should be divided specifically. The typical member of the group is the common dolphin, *Delphinus delphis* (Fig. 6.11). This is found worldwide in temperate and tropical waters, both coastal and offshore. It is one of the most brightly marked of all the cetacea. It is black, or a very dark brown, above, with a white belly; there

Fig. 6.11   The common dolphin, *Delphinus delphis*. The complex pattern of the common dolphin is seen in a modified form in some of the other true dolphins, such as the striped dolphin.

is a complicated pattern of alternating light and dark bands on the flanks, and two waves of yellow and white intersect at the level of the dorsal fin. There is a dark circle round the eye and a tapering band from the base of the flipper to the side of the lower jaw. There is the typical elongated dolphin beak, clearly separated by a groove at its base. There is a slender, sickle-shaped dorsal fin and long tapering flippers. The common dolphin grows to about 2.4 m (8 ft) in length, males being slightly longer than females. This is one of the world's commonest dolphins and must be one of the commonest large mammals. It was the subject of a fishery in the Black Sea, but this has now ceased. Today its main threat is entanglement, either in gill nets, or in tuna purse seines.

A common dolphin, *Delphinus delphis*. The dolphin breathes as it porpoises out of the water – notice the open blowhole.

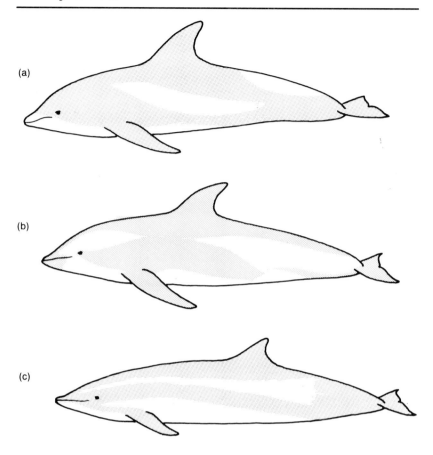

(a)

(b)

(c)

Fig. 6.12   Dolphins: (a) white beaked, *Lagenorhynchus albirostris*; (b) hourglass, *L. cruciger*; (c) Fraser's, *Lagenodelphis hosei*.

The genus *Lagenorhynchus*

Six rather similar dolphins are grouped in the genus *Lagenorhynchus* (Fig. 6.12). The white-beaked dolphin, *L. albirostris*, is found throughout the North Atlantic, from Greenland south to Massachusetts and France. There are 22–25 pairs of teeth in each jaw. The beak, throat and belly are white, while the forehead and back are dark. It grows to about 3.1 m (10 ft). The Atlantic white-sided dolphin, *L. acutus*, is another dweller in the North Atlantic. There are 30–40 pairs of very slender teeth in both jaws. The pigmentation on the back does not extend so far down as in the white-beaked dolphin, hence the 'white-sided' part of its name. The Pacific white-sided dolphin, *L. obliquidens*, is a similar form from the Pacific.

The dusky dolphin, *L. obscurus*, is an inshore species with a circumpolar distribution in the Southern Hemisphere around South America, South Africa, Kerguelen, southern Australia and New Zealand. The hourglass

Pacific white-sided dolphin, *Lagenorhynchus obliquidens*.

Bottlenose dolphins, *Tursiops truncatus*.

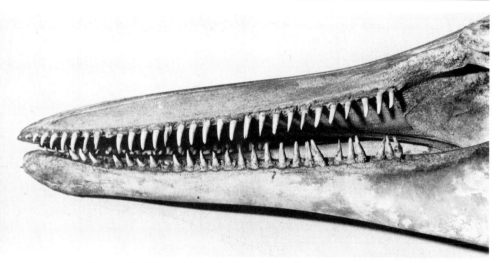

The needle-like teeth of a dusky dolphin, *Lagenorhynchus obscurus*. A dentition composed of essentially similar teeth ('homodont') is characteristic of toothed whales.

dolphin, *L. cruciger*, has a similar range, but appears to be more pelagic. Peale's dolphin, *L. australis*, is another coastal species from South America and the Falkland Islands.

Fraser's dolphin, *Lagenodelphis hosei*, was, until 1970, known from only a single skeleton from Sarawak. Since then, however, it has been reported from the eastern and central Pacific, Japan, Taiwan, eastern Australia and South Africa. It has a short, stubby beak and is regarded as a link between the common dolphin and the *Langenorhynchus* dolphins.

The bottlenose dolphin

The bottlenose dolphin, *Tursiops truncatus* (Fig. 6.14), is a very familiar form, both in the wild where members of this species have, from time to time, developed social interactions with humans, and in captivity, for this is one of the favourite dolphins for keeping in captivity, as it is easily trained and survives well. Much of the work on cetacean physiology and behaviour has been done on this species.

Bottlenose dolphins have a very distinct beak, from which they get their name, and 22–25 pairs of stout teeth in each jaw. These dolphins are coloured slaty-grey or dark brown above, and paler beneath, but there is considerable colour variation in the range, which is practically cosmopolitan, apart from polar seas. They reach a length of about 3.7 m (12 ft), although different populations grow to different sizes. Bottlenose dolphins are usually sociable animals, occurring in herds of up to 40, although, occasionally, herds of several hundred are observed. Sexual maturity is reached in ten to twelve years in males (which are the larger sex) and five to twelve years in females. Gestation lasts a year and is

A frolicking school of Atlantic white-sided dolphins, *Lagenorhynchus acutus*.

followed by a lactation period of 12–18 months. A female bottlenose might have eight calves in her lifetime, separated by two- to three-year intervals. The lifespan is believed to be 35 years or more.

The genus *Stenella*

The dolphins in the genus *Stenella* comprise two clearly defined species and a number of forms which are difficult to distinguish, but together probably make up another four species. The striped dolphin, *Stenella coeruleoalba* (Fig. 6.13), is found in most warm and tropical seas throughout the world. It is a small dolphin, about 2.5 m (8 ft), with a prominent beak, which resembles the common dolphin, apart from the lack of yellow on its sides. It gets its name from the characteristic narrow stripe from the eye along the side. It is heavily exploited in drive-fisheries for food in Japan.

The spinner dolphins are from the tropical waters of the Atlantic, Indian and Pacific Oceans, where there are separate populations. They get their name from their habit of leaping out of the water and spinning around on their long axes before falling back into the sea. There are at least two species, *S. longirostris*, the long-snouted spinner, and *S. clymene*, the clymene

Striped dolphin, *Stenella coeruleoalba*, a warm-water species found in most warm and tropical seas throughout the world.

dolphin. Both are very similar, reaching about 2.1 m (6 ft 10 in) in length, but the clymene dolphin has fewer teeth (about 38–49 pairs in each jaw) than the spinner (45–65 pairs).

Spotted dolphins, to which the names *Stenella attenuata, S. frontalis, S. plagiodon* and *S. dubia* have been applied, are found in nearly all the tropical seas and in some warm temperate regions of the North and South Atlantic also. They have long conspicuous beaks and, as their name implies, are patterned with pale elongated spots on a darker ground, a pattern reversed beneath. The distinctions between the various described forms are minor and relate largely to distribution. We know very little of the genetic isolation of dolphin populations and thus it is difficult to say whether the named forms are really species in the sense that crosses between them would be infertile. Spotted dolphins are surface-feeders, in contrast to the spinners, which are believed to dive deeply for their food. Spotted dolphins are very abundant; their population in the eastern tropical Pacific alone is estimated at 3.5 million. Both spotted dolphins and spinners suffer considerable mortality in tuna seine nets.

Fig. 6.13   The striped dolphin, *Stenella coeruleoalba*.

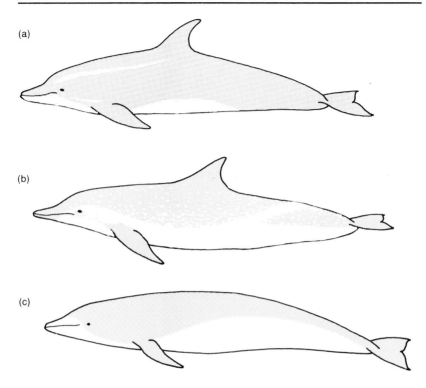

Fig. 6.14   More dolphins: (a) bottlenose, *Tursiops truncatus*; (b) spotted, *Stenella attenuata*; (c) southern right whale dolphin, *Lissodelphis peronii*.

Spotted dolphins, *Stenella* sp.

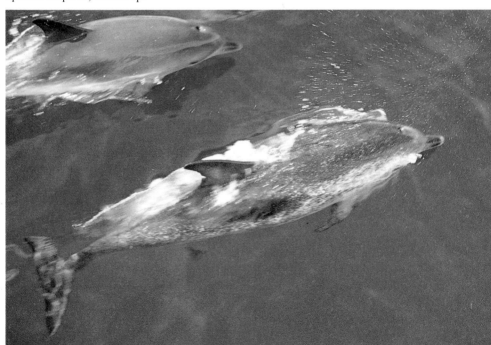

The genus *Lissodelphis*

There are two right whale dolphins, so called because, like right whales, they lack a dorsal fin. *Lissodelphis borealis*, the northern right whale dolphin, occurs in the North Pacific, while the southern right whale dolphin, *L. peronii*, is confined to the subtemperate waters that border the cold Southern Ocean. The northern species grows to about 2.7 m (9 ft); the southern is slightly smaller, around 1.8 m (6 ft). They have slender beaks with about 40 pairs of teeth in each jaw. The upper surface is black, but there is a large white patch on the belly between the flippers and a long narrow streak of white extending from this, around the vent to the root of the flukes. The white of the southern form is much more extensive than in its northern cousin.

The genus *Cephalorhynchus*

The last group of four dolphins to be considered are those of the genus *Cephalorhynchus* (Fig. 6.15). These are all small coastal dolphins, about 1.3–1.6 m (4–5 ft) long, lacking a beak, conspicuously patterned in black and white, and rather poorly known. Heaviside's dolphin, *C. heavisidii*, is from the coastal waters of south-western Africa. Alone in this group, it has a pointed triangular dorsal fin. It is believed to feed on squid and bottom-dwelling fish. The black dolphin (which actually has a white belly), *C. eutropia*, is found exclusively off the coasts of Chile, from Concepcion to Isla Navarino in Tierra del Fuego.

A Commerson's dolphin (*Cephalorhynchus commersonii*) and its seven-day-old calf.

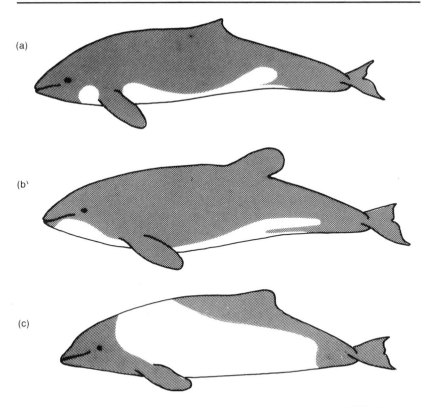

Fig. 6.15 Yet more dolphins: (a) Heaviside's, *Cephalorhynchus heavisidii*; (b) Hector's, *C. hectori*; (c) Commerson's, *C. commersonii*.

Hector's dolphin, *C. hectori*, comes from coastal waters around New Zealand, and is frequently found in turbid estuarine waters. It is usually seen in groups of three to ten, but larger herds of up to several hundred have been reported. Commerson's dolphin, *C. commersonii*, is my favourite of this group. It is found around the southern tip of South America, particularly off Tierra del Fuego and the Falklands. There are also records from Kerguelen, and an erroneous one from South Georgia where the water is undoubtedly too cold for this dolphin. Commerson's dolphin has more white on it than the other *Cephalorhynchus* species. It feeds on small fish, squid and shrimps.

For many years my arrivals at Port Stanley in the Falkland Islands have been greeted by this delightful little dolphin, swimming right up to the jetty and scaring the logger ducks. Sadly, it seems a good deal less abundant nowadays, although there is no evidence of persecution there. It is a different matter in the channels of Tierra del Fuego, where the king crab fishermen catch and kill not only Commerson's, but Peale's, southern right whale dolphins and Burmeister's porpoises as well, to use for bait in their crab pots.

Chapter 7
Killers and Their Kin

The remaining seven members of the family of true dolphins include some of the most impressive and some of the most puzzling of the group. Like the cephalorhynchine dolphins they all lack a beak, but otherwise have few obvious features in common.

The killer whale

The largest, certainly the most impressive, and probably the most familiar, is the killer whale, *Orcinus orca* (Fig. 7.1). This is the largest of all the dolphins, adult males reaching a length of up to 9.5 m (nearly 31 ft), though 8.2 m (26 ft 8 in) is more usual. Females are considerably smaller, around 7.0 m (22 ft 9 in), another example of sexual dimorphism as found

Fig. 7.1   Killer whale, *Orcinus orca.*

in sperm whales. Not only are the sexes different in length; they are different in the porportions of their dorsal fins and flippers also. These appendages grow porportionately faster than the rest of the body; in the case of the flippers, for example, increasing from about one-ninth of the body length in a young male to about one-fifth in an old one. More conspicuous is the dorsal fin, which becomes greatly enlarged in old males, reaching a height of 1.8 m (5 ft 10 in). These huge blades make it easy to locate the presence of the dominant male in a group of killers, for these are highly social whales, travelling about in discrete social units, or pods.

Killer whales are easily identified at sea, for they wear a very distinctive livery. They are startlingly black and white, with a grey saddle behind the dorsal fin. The white is mostly on the belly, with a lobe ascending the flank on each side and another pair of white lens-shaped lobes behind the eyes. The black is jet-black and the white brilliant; only the grey saddle shows any gradation of tone. The flukes are black above and white below, while the rounded flippers are black all over. The mouth of the killer whale is

The startlingly black and white head of a killer whale, as it inspects the world above the water.

Killer whales, *Orcinus orca*, at sea.

A bull killer whale, *Orcinus orca*, in the Antarctic. The high, blade-like dorsal fin is diagnostic of the adult males.

armed with a fearsome array of about 10–13 pairs of stout conical teeth, which in old specimens may be very worn, even to the extent of exposing the pulp cavity. The teeth, up to 5 cm (2 in) of which protude from the gum, interlock; they do not occlude.

Killer whales are found in all the oceans of the world, but mostly in cooler waters. In the Antarctic they are at home amid the pack-ice, but are said not to extend beyond the ice-line in the Arctic Ocean. They are oceanic dolphins for the most part, but will approach the shore when attracted by food resources. In one part of their range, Puget Sound, off the coast of Washington and British Columbia, there is a resident coastal population of about 210 killer whales in some 20 pods, which appear to have a fairly restricted range.

These conspicuous and easily approached whales have provided a unique opportunity for the study of the social structure of the large dolphins. Their dorsal fins have sufficient individuality for painstaking workers to assemble photographic files which allow individuals to be identified, and the composition of the pods followed. One group, the famous 'J Pod' of Puget Sound, can probably claim to be the most studied group of whales in the world. The relationships within the pod are

remarkably stable, and may persist from one generation to the next. Each pod contains one adult male, several adult breeding females and a number of sub-adults of both sexes. The pod size may vary from as few as four to as many as 40, but small pods, which are split off from larger ones, are less stable and may die out within a generation. The larger pods probably monopolise preferred food resources, leaving the smaller ones to make do with what they can find.

It is joint hunting behaviour that accounts for the evolution of the social links that result in pod formation. Killer whales are versatile predators, feeding on a wide variety of prey species, from fish to warm-blooded prey such as birds, seals or even other whales. They are said to require 2.5 to 5 per cent of their body weight of food each day; for a 5,000-kg (11,023 lb) male this might amount to as much as 250 kg (551 lb), so access to a substantial and reliable food source is necessary.

Fish and squid probably form the bulk of their diet, with some pods specialising in warm-blooded prey. We know nothing about their methods of feeding on squid (except for an indication that they may dive deep to do so, for the carcase of a killer whale was found entangled with a submarine telephone cable at 1,030 m [3,378 ft] off Vancouver Island), but killers have been described combining to trap a school of Pacific pink salmon against a rock-face where they could easily be captured. The killers approached the salmon swimming slowly near the surface, vocalising and slapping the surface with their flukes and flippers. Once the fish were trapped, the whales picked them off, one by one, crunching them briefly in their jaws before swallowing them.

This is not, perhaps, so remarkable. Both fish and whales were in the same medium and the entrapment of the former was fairly straightforward. Much more remarkable is the account of killer whale co-operative hunting strategy observed by four biologists in the Antarctic. While sailing through the Gerlache Strait, just to the west of the Antarctic Peninsula, a pod of seven killer whales was sighted. The whales were moving slowly through the pack-ice and occasionally rising vertically out of the water, 'spy hopping', to look at the surface of the ice floes for prey. A lone crabeater seal was lying on a medium-sized floe, about 5–7 m (16 ft 3 in–22 ft 9 in) in diameter. At first one whale approached close to the edge of the floe, rose vertically out of the water and inspected the seal. For the next few minutes the entire pod circled the floe, looking at the seal. The whales then withdrew to a short distance, and returned swimming fast, in echelon. As they neared the floe, they surfaced, coming well out of the water, and then dived quickly, throwing up a large wave that washed over the floe, tipping it up at a steep angle, and throwing the seal into the water.

Although the whales were not observed to catch and eat the seal, the biologists were convinced that what they had seen was purposeful action. The whales inspected the situation, devised a suitable plan and carried it out in unison. Such actions are more indicative of higher mental powers than any amount of training in research institutes or dolphinaria.

Herbert Ponting, the photographer with Scott's expedition to the South

Killer whales, *Orcinus orca*, 'spy-hopping'. This group of killers is eyeing the photographer at the edge of the ice. Killers are known to size up the chances of tipping seals into the water from such a position.

Pole in 1911, recorded a chilling experience. He was standing at the edge of the sea-ice on the Ross Sea, photographing a pod of killers cruising up and down the ice-edge:

> 'I had got to within six feet of the edge of the ice – which was about a yard thick – when, to my consternation, it suddenly heaved up under my feet and split into fragments around me; whilst the eight whales, lined up side by side and almost touching each other, burst from under the ice and "spouted".'

Ponting was convinced that the whales had eyed him as a potential dinner. It may have been simply curiosity, but the episode with the crabeater seal is suggestive.

As remarkable are the accounts of killer whales taking southern sealions from their breeding beaches at Punta Norte, Chubut, in Argentina. The whales came right to the edge of the surf, almost beaching themselves, in order to snatch the sea lions; a remarkable display in a cetacean that seems susceptible to stranding in the world!

There are many stories of killer whales feeding on other whales. The first of these originated from the whale fishing off the coast of New England and was recounted by the Hon. Paul Dudley (the same man associated with the 'scrag whale') in 1725. He told of a pack of ravenous killers belabouring a wretched right whale, catching hold of its lips until the tongue lolled out, and then devouring this morsel. This does not seem to have been observed since, but we should not dismiss it as fantasy. Killer whales definitely have a liking for the tongues of rorquals. During the heyday of Antarctic whaling, when it was commonplace to have a dozen or two blue and fin whales moored to the stern of a factory ship awaiting processing, a man would be stationed on the poop deck with a rifle to shoot at the killers, which would otherwise come up to eat the tongues. The tongue of a fin whale would produce nearly a tonne of oil and the whalers were reluctant to let the killers take this.

From Twofold Bay, in New South Wales, comes another story of an association of men and killers. Twofold Bay lies close to a migration route of humpbacks and right whales. A shore whaling station was set up there in 1818 and whaling continued there till 1929 (or perhaps 1932). Regularly, in July each year, killer whales would appear in Twofold Bay. According to the whalers, these whales would station themselves in three groups, waiting for passing rights or humpbacks. If one group detected a whale, they would attract the attention of the other groups by lob-tailing (or perhaps by vocalising?) and then assemble to mob and worry the whale, and perhaps kill and eat it.

A pair of killer whales, *Orcinus orca*, are put through their paces at an dolphinarium. Such performances do little for the dignity of whales, but they do help to make the public sympathetic towards them.

A killer whale, *Orcinus orca*, curves through the water by the carcase of a sei whale, *Balaenoptera borealis*, left 'in flag' by a whaler. The carcase has been inflated with air so that it will float, and marked with a flag and a radar reflector to aid recovery. The killer whale will eat the tongue from the dead sei.

The whalers took advantage of the obvious signals from the killers to launch their boats after the whale, which the killers prevented from escaping to sea. Once the whale was harpooned the whalers reported that regularly four of the killers would separate from the rest and while two of them took up station underneath the head of the whale, so as to prevent it from sounding, the others would throw themselves over the blowhole of their quarry, so as to hinder its breathing. This rendered the eventual lancing of the whale by the whalers a far easier task, for which the killers were rewarded with the tongue of the dead whale, which, as noted earlier, was for them a delicacy.

This association of commensal fishing was recorded by Professor W. J. Dakin, Professor of Zoology at the University of Sydney, who described it in his book *Whalemen Adventurers*. It is a remarkable tale, which Dakin described as intelligent co-operation. It is perhaps more reasonable to regard it as a rather complex training routine, taking place with free-living dolphins rewarded by their share in the dead whale.

Scammon reported killer whales eating greys and, more recently, this was confirmed when a pack of killers was seen to kill and eat a young grey whale off California. Off Cabo San Lucas, Baja California, in Mexico, a pod of killers corralled a young blue whale for hours and held it at the surface while individuals charged in to tear away chunks of flesh. The incident ended when the killers abandoned the badly injured blue whale.

The killer whale is clearly an awesome predator, and one which humans would do well to respect. These are no records of human deaths attributed to killer whales, although several yachts claim to have been attacked by killers. In some cases the identification of the whales seems a little doubtful; the schooner *Lucette* was rammed and sunk by a group of about 20 whales in 1972 and these probably were killers. Why they should have done this is quite unknown. Perhaps the yacht had sailed between a mother and her calf, or rammed a whale. We shall never know.

What is quite certain is that killer whales are the gentlest of creatures in captivity. An impatient killer may give its trainer a shove, or even grip his arm in its mouth, but no serious accidents are reported, although it would be very easy for a killer whale to kill its trainer if it determined to do so. Killers are universally loved by those who care for them and by the public that come to watch them. Because the name 'killer' has a perjorative sound to it (we sometimes forget that nearly all meateaters are killers and that humans are the greatest killers of all), there has been a move to refer to these dolphins as orcas. I am generally opposed to changing the names of animals to suit current, and probably transitory, fashion, but orca is derived from the second part of the killer's scientific name, *Orcinus orca*, and has a respectable pedigree in English usage, although not in America until this recent trend. The Norwegian whalers called them '*spekkhogger*', or 'blubber-chopper'. It seemed a good name to anyone who had watched a pack of killer whales tearing into the carcase of a shot whale, left 'in flag' to be picked up later!

False killer whale

The false killer whale, *Pseudorca crassidens*, as its name suggests, is a relative of the killer. It is a large dolphin, reaching 4–5.5 m (13 ft–17 ft 11 in), with the males larger than the females, although not to the same extent as in its larger cousin. It is rather slender in outline with a thin, somewhat blunt, pointed head (Fig. 7.2). False killers are black all over, apart from some white speckling on the leading edge of the flippers and a greyish area on the belly between them. The flippers are narrower than those of the killer whale, and, strangely, have a distinct and obvious hump half way down the leading edge of each. The dorsal fin is only up to 40 cm (16 in) high, and nothing like the great blade of the bull killer. There are nine to eleven pairs of stout teeth in each jaw, which are circular in cross-section,

Fig. 7.2   False killer whale, *Pseudorca crassidens*.

The false killer whale, *Pseudorca crassidens*.

and can be distinguished from those of the killer whale, whose teeth are oval when cut across.

False killer whales are found worldwide in tropical and subtropical waters, and sometimes extend into cooler areas. Up to 1927 the only specimens known from the British Isles were three subfossil skeletons from the fens, but in that year a school of 150 stranded in the Dornoch Firth, and since that time there have been several other strandings. Indeed, it seems to be one of the characteristics of this large dolphin that it tends to run ashore *en masse*, as many as 300 animals perishing in the bigger incidents.

False killers feed on fish and squid, and probably tend to take large prey. They have been seen with fish estimated to weigh 9.1 kg (20 lb) in their mouths, shaking their prey to dismember it. A sub-adult false killer, kept in captivity, consumed, a daily average of 20.4 kg (45 lb) of fish and squid, amounting to 4.6 per cent of her body weight.

Pygmy killer whale

A much smaller dolphin, the pygmy killer whale, *Feresa attenuata* (Fig. 7.3) is a third member of this group. These are tropical dolphins from the Pacific, Atlantic and Indian Oceans. Until 1950 the species was known only from a couple of skulls, but since then several stranded specimens have turned up and it has also been observed in the wild and taken into

Fig. 7.3   Pygmy killer whale, *Feresa attenuata*.

captivity. It seems to be an aggressive little dolphin whose presence evokes fear in other cetaceans in the same tank. One specimen kept at the Sea Life Park in Hawaii behaved aggressively from the moment of its capture, snapping at anyone who came within range. Eventually it grew more tolerant of its attendants, but still would not allow close approach, a very different pattern of behaviour from that of most other dolphins, which actively seek out contact with their trainers.

The pygmy killer whale reaches 2.1–2.4 m (7–8 ft) and a weight of about 160 kg (350 lb). Most of its body is black or dark grey, with a paler area like a cape over the back. There is a pale, anchor-shaped patch on the chest between the flippers and usually some white on the belly. Amazingly, it is reported to be able to extrude its eyes from their sockets sufficiently far to be able to look behind itself! I have never seen a live pygmy killer whale, but if I do get the chance I shall look out for this feature.

Melon-headed whale

Rather similar to the pygmy killer is the clumsily named melon-headed whale, *Peponocephala electra* (Fig. 7.4). Its vernacular name is a direct translation of its generic name, and I much prefer to call it the electra dolphin, after its specific name, but this term does not appear to be in common use. This is another warm-water dolphin which was once thought to be extremely rare but, as in the case of the pygmy killer whale, there

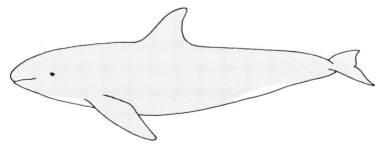

Fig. 7.4   Electra dolphin, or melon-headed whale, *Peponocephala electra*.

have been some large recent strandings. One need not suppose that these strandings are a new phenomenon; almost certainly they have been occurring at intervals ever since the dolphins evolved. What has changed is human interest in whales, which has led to better reporting of these incidents, so that specimens find their way to museums.

Very little is known of the electra dolphin. It is a gregarious species, occurring in herds of several hundred, but although it will occasionally indulge in bow-wave riding, it is generally difficult to approach.

Risso's dolphin

Risso's dolphin, or grampus, *Grampus griseus*, (Fig. 7.5) is another member of this group. This is a large dolphin reaching 4.3 m (14 ft) and 680 kg (1,500 lb). There is no evidence of a difference of size between the sexes, and we might deduce from this that there is none of the complex social organisation and male polygyny in this species that is found in the sperm or killer whales. The body is more robust than the three species described earlier, with long, tapering flippers and a large dorsal fin set rather far back. The head is bulbous, with a V-shaped groove at the front. There are no teeth in the upper jaw and only four pairs near the front of the lower. Risso's dolphin is uniformly light grey at birth, but during the first few years the skin darkens almost to black before again beginning to lighten. By the time the animal is adult it is largely creamy white or silver grey except for the dorsal fin and the adjacent part of the back, the flukes and the tips of the flippers, all of which remain pale buff to dark ochre. The body is usually covered with white oval scars and scratch marks. These may be caused by interspecific fighting or in tussles with the squid on which Risso's dolphins feed almost exclusively.

The species has achieved modest fame through the exploits of 'Pelorus Jack', a Risso's dolphin that used to escort ships crossing Cook Strait between the two main islands of New Zealand from 1888 to 1912. He (or it may have been she) would do this over a certain stretch of water and nowhere else. A south-bound steamer would be joined near the entrance to Pelorus Sound and accompanied as far as French Pass, and similarly in the reverse direction for ships going to Wellington. Pelorus Jack seemed to respond to the sound of the ship's screw and would leave off whatever else

Fig. 7.5   Risso's dolphin, *Grampus griseus*.

he happened to be doing to take up station alongside or ride the bow wave. According to one observer, Jack, if given the choice between two steamers, would always choose the faster one. He could effortlessly keep up with a vessel travelling at 27 km/hr (15 knots).

Inevitably, some mindless vandal shot at the dolphin with a rifle. As far as is known, this did no harm, but the act did stimulate the Governor of New Zealand, Lord Plunket, to sign, on 6 September 1904, an Order in Council prohibiting the taking of Risso's dolphins in the waters of Cook Strait. In 1912, when Pelorus Jack had been escorting vessels for nearly 24 years, he went missing from his station and was seen no more.

The reason for this kind of behaviour is unknown. The ships offered the dolphin nothing as a reward for following, unless it was the satisfaction of riding in the bow wave. Perhaps Perlorus Jack lacked company – Risso's dolphins are usually found in groups of a dozen or so, though solitary animals are not uncommon – and the passing ships satisfied this need. He was reported to rub himself against the ship's side, and this may have been a pleasurable sensation. It is, in the end, unexplained behaviour.

The genus *Globicephala*

The remaining whales in this group are the pilot whales, or pot-headed whales (Fig. 7.6). Two species are recognised: the long-finned pilot whale, *Globicephala melaena*, and the short-finned pilot whale, *G. macrorhynchus*, which differ in the characteristics that their names suggest. Their body

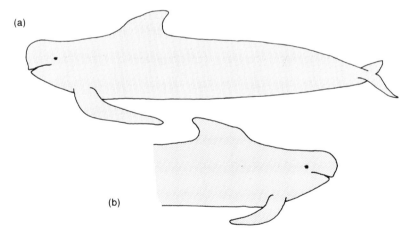

Fig. 7.6 Pilot whales: (a) the long-finned pilot whale, *Globicephala melaena*; (b) the short-finned pilot whale, *G. macrorhynchus*

colour is black (they were often called blackfish by the old whalers, who valued them for their oil when they could get nothing better), or very dark brown. Sometimes there is a grey saddle behind the prominent rounded dorsal fin, which is set rather far forward, and there may be a

A stranded school of long-finned pilot whales, *Globicephala melaena*. The reasons for such strandings are still a mystery.

pale anchor-shaped mark on the chin. The bulbous head is very characteristic; in old males the 'forehead' may be so expanded as to overhang the upper jaw. There are eight to ten pairs of slender teeth near the front in both jaws. They are large dolphins, though with a rather slender body shape.

The long-finned pilot whale reaches a length of 6.2 m (20 ft 4 in) or even as much as 8.5 m (27 ft 11 in) in males; females are smaller, reaching about 5.4 m (17 ft 9 in). Their elongated sickle-shaped flippers may be up to one-fifth of the total body length. They live in the cooler waters of both hemispheres and the northern and southern populations are genetically distinct. The short-finned pilot whale is slightly smaller, males reaching 5 m (16 ft 5 in), or, exceptionally, 7 m (23 ft), and females 3.6 m (11 ft 10 in). They have a more tropical distribution in the warmer waters between the two populations of the long-finned species.

The size difference in the sexes suggests that there is some competition between the males for the females, and a degree of polygyny. Indeed, it was found that in summer herds of long-finned pilot whales, investigated off Newfoundland, over twice as many adult females as males were present. As we can safely assume that, as in all other mammals, the sexes are born in about equal numbers, this implies that either about half the males had left the herds, that there had been a much greater mortality of males, or that there is a different age of sexual maturity for males and females. The answer is probably a combination of the last two explanations. Male polygamous mammals usually have a shorter lifespan than females and male long-finned pilot whales are not sexually mature till they are 11 years old, while females become mature at six or seven years old. Adult male pilot whales are often heavily scarred and have been observed biting at each other, butting with their heads and fluke-slapping. These contests probably serve to set up a dominance hierarchy among the males, for access to the females.

In behaviour long-finned and short-finned pilot whales are very similar. They are highly gregarious, travelling in huge bands, often of many hundreds. They are frequently found in the company of other cetaceans, particularly bottlenose dolphins. When travelling, herds of pilot whales appear to be highly organised. The herd is split into smaller bands, which move in long lines of whales all abreast and separated by a few metres or tens of metres. It is believed that this configuration increases the prey-detection ability of the herd as a whole. When prey is found, the structure of the group breaks down as individual whales pursue their food. Pilot whales are very inclined to loaf about at the surface, particularly in the early or mid-morning, presumably after a night's feeding. From what we know of their feeding habits, they subsist chiefly on squid, although shoaling fish are also taken.

Pilot whales are very susceptible to mass stranding. The reasons for this are unknown, but some biologists believe that large herds may strand sufficiently frequently for this to have a significant effect on populations. Pilot whales are reasonably abundant dolphins – the population in the North Atlantic is probably in the high tens, or even low hundreds, of thousands.

The combination of abundance and the tendency to strand probably brought primitive people into early contact with pilot whales. The happy discovery of a pod of stranded whales on the shore would ensure a huge supply of meat, and, perhaps more important, oil, for the wandering bands of subsistence hunters that patrolled the northern shores in the Stone Age. Bones of whales are commonly found at prehistoric sites, and sometimes in association with stone adzes, which might have been used to strip the blubber from the carcase. The blubber oil would have been

Long-finned pilot whale, *Globicephala melaena*.

valued not only for its edible qualities, but also for its ability to provide light and heat when burnt in stone lamps. It is easy to see how useful such a fuel would be through the long northern winters.

The scavenging of beached whales might have led on to the hunters taking measures to ensure that passing pods of whales conveniently stranded on accessible beaches. Pilot whales are very easily driven and once one whale has gone ashore, the others will follow. Such 'drive fisheries', as this method of hunting is called, developed in many parts of the northern world. One of the oldest of which we have historical records, and one that has persisted until the present day, is the drive fishery in the Faeroe Islands.

The pilot whale, or '*grindhval*' as it is known in the Faeroes, plays a more important part in the islands' culture than it does in their economy, and the whale hunt, or '*grind*', is pursued with great eagerness. When a school of pilot whales is sighted, boats are launched to shepherd the whales into a chosen killing place where they can be driven ashore and from where the meat can be easily carried away. The whales are driven in front of the boats by stones tied to short lines being thrown into the water, or by a disturbance in the water being created with oars or paddles. It is said that the whales will not be driven against the tide, but like to swim against the wind. The vital thing for the hunters is not to let the herd turn under the boats and escape beneath them.

Once the whales are in shallow water and the first few have come ashore, the success of the *grind* is assured. The beached whales are lanced or killed by stabbing them in the neck, through the spinal cord. One of the poignant details that figures in many accounts of *grinds* is the frenzied activities of the men, wading about in the blood-stained water up to their waists, as they go about their bloody business of killing the whales, which contrasts strangely with the calm of the whales, lying still even as the knives and lances are plunged into them.

Once the killing is over, the division of the spoils begins. This is done on a traditional basis, following practice handed down from long ago, and supervised by a specially appointed offical. The largest whale (officially measured by another functionary) is allotted to the man who sighted the school; other whales are set aside to provide meat for all the houses in the settlement where the kill was made; while yet others are used to provide compensation for any damage to boats or gear during the *grind*.

Even recently, most of the meat was used as food. The practice was to boil it with potatoes and mustard sauce. The liver, heart, kidney and tongue were also used. Blubber was served with the meat, or rendered down for oil. Meat might be salted away for the winter, or hung up in strips to dry in the wind.

A successful *grind* was followed by an evening of dancing and socialising. It is easy to see how, in the past, the killing of a school of pilot whales could save a settlement, eeking out a precarious living from sheep farming and fishing, from the spectre of hunger during the winter. Times have changed in the Faeroes, however, and the economy is now relatively buoyant. But these island peoples cling tenaciously to their traditions. It is

A small school of long-finned pilot whales, *Globicephala melaena*. The round head ('pot head') can be clearly seen.

hard to see very much difference between what goes on in the Faeroes and the subsistence whaling in Alaska, except that the Faeroes' catch is not made up of an endangered species.

Records of the Faeroese drive fishery for pilot whales have been kept since 1584 (and the fishery itself goes back to much earlier times). In the first 300 years catches averaged less than 400 a year (probably because of the low population of the islands). Recently, catches have been around 1,500 a year and it seems that a catch of this size is sustainable, although there are fluctuations in the size of the pilot whale stock. Sighting surveys seem to indicate that long-finned pilot whales in the north-east Atlantic have actually been increasing in recent years, perhaps as part of a long-term cycle. In Newfoundland, where a drive fishery was active between 1947 and 1964, with catches occasionally reaching 10,000 a year, there was a significant impact on the populations. Short-finned pilot whales are hunted in Japan today, in Okinawa and on the Izu Peninsula, and a few hundred are taken by hand-harpoon off the island of St Vincent in the Lesser Antilles.

Pilot whales are easily trained and seem to survive better in captivity than many other cetaceans. They are attractive animals, which, while seeming to maintain their independence of their trainer (they respond much better to rewards than to punishment), still elicit our sympathy and affection. We ask ourselves how can the intelligent, well-educated Faeroese still indulge in the needless slaughter of these harmless animals? No clear answer comes back to us. Distasteful as it may seem to people who live in cities, or removed from the more basic contacts with nature, there is, for many people, a deep satisfaction in the hunt and the carrying back of food to the home. Perhaps Faeroese pilot whalers or Alaskan bowhead hunters are not killing for food, or even for culture, but to satisfy an age-old instinct.

Chapter 8
Narwhals and Belugas

The cold waters of the Arctic are the home of two toothed whales, the narwhal, *Monodon monoceros*, and the beluga, or white whale, *Delphinapterus leucas*. The narwhal, or sea-unicorn, is the possessor of what is probably the strangest tooth in the animal kingdom.

The narwhal

Projecting from the left side of the head of the male narwhal (Fig. 8.1) is a massive ivory tusk, up to 3 m (9 ft 10 in) in length, or three-fifths of the body length of the whale. The tusk is marked with a spiral twist, like an old-fashioned stick of barley-sugar, spiralling to the left, or anticlockwise, from the whale's point of view, and sometimes there is also another spiral, like a corkscrew, superimposed on this.

Narwhal tusks were familiar to the civilised world long before the animal that grew them was known. Like all ivory, narwhal tusks were highly prized, but without certain knowledge of their provenance it was easy to persuade the gullible that they were the genuine tusks of the mythical unicorn, and thus endowed with magical powers. A particularly useful feature of cups made from unicorn ivory was that they would neutralise any poison placed within them. In a world of superstition some pragmatism survived, however; Mukarrab Khan declined to purchase a unicorn's horn after testing its effect, with disappointing results, on poison offered to a pigeon, a goat and a man.

Throughout the Middle Ages, and later, a flourishing trade brought narwhal ivory from the Arctic to Europe, much of it for further shipment to the East. Viking traders, from their Greenland colonies, seem to have been the first to bring back narwhal tusks, but when the Viking settlements died out, so too did the trade. The right whalers, from the seventeenth century onward, took the occasional narwhal, and also bartered with Eskimos for 'unicorn horns', although by this time the magical properties were of less appeal than the intrinsic beauty of the ivory spiral. However, even in the eighteenth century, the Japanese valued narwhal tusks for their powers to prolong life, fortify the animal spirits,

Fig. 8.1   A male narwhal, *Monodon monoceros*.

146

assist the memory and cure all complaints. Powdered narwhal ivory was sold as the drug '*ikkaku*' in Japan as recently as the 1950s.

What is the origin of these tusks? In the foetal narwhal there are two pairs of tooth germs at the front of the upper jaw, behind which are four pairs of dental papillae, which may occasionally develop into tiny teeth concealed beneath the gum. Of the anterior tooth germs, the posterior pair degenerate but the front pair develop further and, in males, one of these, the left, grows out as the tusk. This tooth continues to grow throughout the life of its bearer, the pulp cavity remaining open and new dentine being deposited continuously. The right tooth grows to about 2.5 cm (1 in), but remains embedded in the gum. In female narwhals both front teeth develop like the right tooth of the males. Very occasionally both teeth will develop in males, producing a symmetrical pair of horns. Equally rarely, a female will produce a tusk.

Biologists have puzzled over the function of the tusks for almost as long as the narwhal has been known to science. Most of the speculation has been very theoretical because the narwhal's high Arctic habitat makes direct observation difficult. It has been suggested that the tusk is used as a weapon, as a means of spearing food, to stir up sediment to disturb prey, or as a tool with which to break ice. Two more recent suggestions have been that the tusk functions as a heat-exchanger, transferring excess heat from its well-vascularised pulp cavity, or that it is a sort of wave-guide for sound production, vocalisations produced in the head being channelled down the tusk to emerge at maximum intensity at its tip.

It does not follow, of course, that the tusk has only one function, as an organ evolved for one purpose may be employed for others, but it is likely that there is one primary function to be identified. It is significant that the tusk is found exclusively (or almost exclusively) in males. This suggests that any function that would apply equally to females is unlikely to be the primary purpose of the tusk. Therefore, unless female narwhals take very different food from males, and there is no evidence that they do, it is improbable that the tusk is used for food gathering. Similarly, it would seem that female narwhals would have just as much need to break ice as males. Had the tusk been developed as an ice-breaker the females would be as well equipped as the males. Male narwhals are slightly larger than females, growing to about 4.7 m (15 ft 5 in), excluding the tusk, while females rarely exceed 4.5 m (14 ft 9 in), so it is unlikely that the males have a significantly greater problem of heat-exchange than the females. In any case, there is no reason to suppose that either sex of narwhal cannot dispose of excess heat using the method described on pages 20–1.

A requirement to channel sound in a different manner from the females might well be important to the male narwhal. There are a great many mammals in which part of the sexual display of the males involves the production of characteristic sounds – the bellowing of a bull, the screams of a howler monkey or the roar of an elephant seal are all examples. But the explanation may be simpler that this. A narwhal's tusk is typical of many of those weapons or adornments that are found in so many male mammals and are used in establishing sexual dominance. Scottish right whalers recorded seeing narwhals fencing with their tusks, and more recently Helen Silverman, an Arctic biologist, spent three seasons in northern Baffin Island studying the behaviour of narwhals; she often saw males crossing tusks with each other. These encounters did not appear very aggressive to her, but the observations were limited to the summer period, from June to October, while the narwhal's breeding season falls between March and May. She inferred that the behaviour she had observed might help to maintain dominance hierarchies and enable young narwhals to develop the skills necessary for the performance of adult sexual roles.

That narwhals do use their tusks aggressively can be inferred from the high incidence of scarring on the heads of adult males and from the occasional finding of pieces of broken tusk embedded in the skull of male narwhals. There are even reports of a broken tusk with the tip of another tusk, jammed in the hollow end, which would seem to indicate a high level of precision in the fencing! There seems little doubt that the tusks are used in fighting to establish the right of preferential access to females, but, as in most mammals, fighting is probably an exceptional last resort after other displays have failed, and the main function of the tusk may be to act as a visual (and perhaps, acoustic) signal to other males. The signal function would account best for the extravagant length of the tusk – as a weapon it must be very unwieldy, to say the least – but if its function is primarily visual, then the bigger the better. It would be very interesting to know how sexually successful the narwhals with two tusks are.

Apart from their tusks, narwhals are not very remarkable whales. As mentioned above, they are small whales, males reaching a weight of about 1.6 tonnes. The head is bluntly rounded, with a rather bulbous forehead. There is no dorsal fin, but the ridge of the back over the posterior third of the body is sharply pointed. The flippers are short and broad and rounded at their tips. The flukes are conventionally shaped in females and young males, but in the older males differential growth causes the tips of the flukes to swing forward, giving a concave leading edge. At birth the young are coloured dark brown or grey, but become progressively lighter as they age, developing a mottling of grey and dark brown flecks on a paler background. This is a pattern which is unknown elsewhere in cetaceans and is instantly characteristic of the narwhal. It has been claimed that the name 'narwhal' is an allusion to this patterning, being derived from an Old Norse word 'nar', a corpse, a reference to the presumed likeness of the narwhal's mottled skin to that of a drowned man. This seems far fetched and fanciful etymology to me, but we have no better suggestion.

Narwhals are seen further north than any other cetacean. Large concentrations are found in Davis Strait, Baffin Bay and in the Greenland Sea. Smaller numbers are present in Hudson Bay, Foxe Basin and the Barents Sea. There are a few thousand in the Soviet Arctic, but the narwhal is rarely found in the Beaufort Sea. There are perhaps between 25,000 and 50,000 narwhals altogether and it is generally believed that the species was once more abundant and more widely distributed.

Narwhals make considerable migratory movements, moving into deep bays and fiords during the summer, and in the winter moving to more oceanic water where there is less chance of a freeze-up trapping and drowning the whales. When the brief Arctic summer ends abruptly, narwhals are sometimes trapped in small pools of water from where they are unable to swim, under possibly miles of ice, to the open sea. Such a situation is referred to by the Greenlandic term '*savssat*'. Natural mortality of narwhals in a *savssat* may be high and the opportunity has historically been taken by Eskimos to kill the whales. The hunters had only to stand by the narrowing pool of open water and wait for the trapped narwhals to come up to breathe, as they must. It was then an easy manner to harpoon or shoot them as they rose. In the winter of 1914–15 at Disko Bay, West Greenland, more than 1,000 narwhals were killed by hunters in this way. Such mortality is probably not additional to that which the narwhals would have suffered anyway, as all the whales would have drowned once the ice froze over completely.

Conception usually occurs about mid-April, with births following a 15-month gestation period, in mid-July. The young are born in sheltered water when the pregnant females enter their summering grounds. The newborn calf weighs about 80 kg (176 lb) and has a 2.5 (1 in) layer of blubber to keep out the cold. The birth of a calf has been recorded as far north as 84°! Initially the young narwhal relies chiefly on plentiful supplies of milk from its mother to make up for energy losses in the form of heat to its environment, but as its blubber layer builds up, so more of its milk can be used for growth. Lactation lasts about 20 months and the likely calving interval is three years.

The narwhal, like many other toothed whales, appears to feed largely on squid, though Arctic cod, Greenland halibut and shrimps of several species are also taken. Murray Newman described how a tethered narwhal at Ellesmere Island would readily take four-horned sculpins from its captor, but it is not known whether they take this bottom-dwelling fish in the wild. The narwhal's small mouth, devoid of functional teeth, looks poorly designed to catch fish or squid. Perhaps it relies mainly on suction to engulf its prey.

The beluga

Narwhals are very vocal whales, producing a wide variety of clicks, squawks and whistles, but they are not as vocal as the next species, the beluga, which was known to the old whalers as the 'sea canary' on account

Fig. 8.2   The white whale, or beluga, *Delphinapterus leucas*.

The beluga or white whale, *Delphinapterus leucas*.

of its squeals and chirps that were easily audible through the hulls of their boats. The complexity of the vocal repertoire of the beluga exceeds that of any other whale and it is impossible not to believe that it has considerable significance in the life of these whales. Bill Schevill and Barbara Lawrence monitored the activity of these whales in the estuary of the Saguenay River in Quebec. They described the calls as high-pitched resonant whistles and squeals, ticks and clucks, mews, chirps, resonant bell-like tones found in no other whale, sounds resembling an echo-sounder, sharp reports and trills, and other sounds suggesting a crowd of children shouting in the distance! It is tempting to suppose that this remarkable range of vocalisations represents a language of considerable complexity, but there are no behavioural observations to support this.

Belugas are the only whales that are truly white (the name '*beluga*' means 'white one' in Russian), and even they are not white for the whole of their lives, for, as in the narwhal, the young are born dark. However, whereas in the narwhal ageing leaves a paler, but still speckled, whale, in the beluga the pigment fades absolutely, to leave a completely white animal. Belugas grow to about 3–5 m (9 ft 10 in–16 ft 5 in), the actual size varying among different populations, but always with the males being slightly larger than the females. There is no dorsal fin, but a shallow ridge,

A small group of belugas, *Delphinapterus leucas.*

A group of narwhals, *Monodon monoceros.* The convex trailing edge of the flukes is clearly shown.

which may be interrupted, runs down the centre of the back (Fig. 8.2). The flippers are narrower and more pointed than those of the narwhal, but show the same tendency to turn up at their tips. There are about eight to eleven pairs of simple conical teeth in the upper jaw and eight or nine pairs in the lower jaw. As the animal ages, these teeth become very worn down and tend to drop out, so that older animals (which may live up 25 or 30 years) have few teeth.

The most remarkable feature of the beluga is the great mobility (for a whale) of its neck. A beluga can turn its head sideways to look behind it and, having done so, can alter its facial expression in a manner also quite unlike any other whale. The bulbous melon that surmounts its head is filled with oil like that of other toothed whales, but belugas seem to be able to adjust the shape of the melon. The function of this is not clear, but it adds to the versatility of visual signalling by markedly altering the outline of the head. Jaw-clapping produces signals that are both audible and visual, and it may be that the teeth, which do not emerge from the gums until the beluga is two or three years old, are more important for their role in an open-mouth display than they are for feeding.

This emphasis on signalling suggests that belugas are highly social animals, and what we know of their habits confirms this. Belugas occur in herds that may number hundreds or even thousands and, in the past, the larger herds may have been commoner. A beluga herd is closely aggregated on the breeding grounds, but more spread out during feeding. Within the herd there is segregation by age and sex. Females form nursery groups with newborn and older calves up to the age of two years, while males group together. Whether the groups of adult males represent non-breeders, excluded from the females by dominant males, or whether they are all potential breeders, is not known. It is not unusual to find male belugas with bites on their flippers and flukes, but there is no evidence of a single dominant harem master accompanying the breeding females, as in the sperm whale.

Beluga calves are usually born between May and July in warm waters from 10–20°C (50–68°F). Such temperatures are to be found in arctic seas only where shallow waters are warmed by the continuous sunlight of high latitudes. In a few localities the calves may be born as early as March or as late as September. The gestation period is about 14 months and lactation lasts for one or two years. During this period the mother and calf maintain very close contact. The nursery pods may contain extended family groups, which appear to feed co-operatively. Several whales of different ages will swim shoulder to shoulder to force a shoal of fish towards a sloping beach where they can be trapped. The whales can then swim into the crowded shoal to snatch the fish.

Belugas feed on a wide range of food organisms. Fishes such as capelin, cod, herring smelt and flounder make up the bulk of their diet in the summer, with polar cod becoming the dominant form in winter. Besides fish, belugas take squid and octopus, crabs, shrimps, clams and even sandworms. In certain estuaries the beluga's appetite for salmon has earned it an evil reputation. In Alaska there has been some success in using underwater recordings of killer whale vocalisations – beluga spookers, they are called – to discourage belugas. If, in fact, beluga spookers do work for more than a few weeks, it does not say much for the level of intelligence of these whales. If they were really perceptive they might be expected to regard the recordings – not being reinforced by the actual presence of killers – to act as a dinner-gong rather than as a deterrent.

Belugas have a wider distribution than narwhals. They are found in Arctic waters off the North American and Soviet coasts and descend to sub-Arctic waters in the Bering Sea and on the Canadian east coast. Their numbers are not known for certain, but are probably between 40,000 and 55,000, although they were certainly far larger than this in the not very distant past. Present hunting rates in eastern Canada, the Barents and White Seas are likely to cause the stock size to decline still further.

Belugas, because of their predictable migrations and coastal breeding, have been the subject of human predation since people first roamed the Arctic shores. Subsistence hunting by Eskimos, prior to the arrival of Europeans, probably had a negligible impact on stocks. Even the arrival of the right whalers in the eighteenth century would have made little difference. It was not until the middle of the nineteenth century, when the whalers began systematically to supply the Eskimos with firearms, that the balance shifted. From small beginnings, in single kills of whales, very destructive net fisheries developed. At Churchill, Hudson Bay, George Simpson McTavish, an employee of the Hudson's Bay Company, wrote:

'Sometimes the whales would not come down with the turn of the tide, and when this happened we were unable to lift the barrier net or gate when the current became too strong. We therefore took a small boat and proceeded with the flood tide to the vicinity of Mosquito Point, and at the proper moment harpooned a whale. This was easily done, as they followed the sailing boat out of curiosity, and when under the stern a swift heave of a hand harpoon transfixed one. The result showed the sound-carrying capacity of the water (over four times that of air), as on this lone whale being struck, all the white whales dived and disappeared simultaneously, and when they reappeared to blow, they were all heading for the mouth of the river at full speed. The net was so arranged, fastened by one end to the shore, and the other anchored in the stream, that there was no wholesale obstruction to the whales, but we got what might be called a good catch, the stragglers following the immediate shore line.'

The oil obtained from white whales was of the same quality as sperm oil and sold for a good price as a lubricant and illuminant. Besides the oil, the hide was of value. This was stripped off the whale, divided into two pieces, scraped and pegged out to dry. After drying, the hides were shipped back to England where they became the 'porpoise hides' of commerce. The meat from the whales was consumed locally by the people, Eskimo and European, at the post.

A beluga fishery existed at Churchill until 1968. The Hudson's Bay operation was closed down in 1931, leaving whale hunting to the local residents and missions, who used the meat for dog food. In 1949 the beluga fishery came under the control of the Federal Department of Fisheries and, shortly after, a commercial company shipped in a factory to process whales. Locals of the district, Eskimo, Cree and Europeans alike, purchased $1 licences to hunt beluga and were paid by the foot for their catches.

The fishery flourished throughout the 1950s, with an average yearly catch of 450 whales. After a short period of decline it was taken over by another company in 1965. Hides for leather, head oil for fine lubrication, steaks for sale in Winnipeg, *muktuk*, (the skin and blubber) for the delectation of the Eskimos, and the carcase oil for margarine were the principal products. The carcase and bones were ground into meal for dog food and mink farm feed. Fortunately for the whales, a decline in the market (largely brought about by concern at the high, but wholly natural, concentrations of mercury in the whale meat, and the closure of many mink farms) resulted in the final closure of the enterprise in 1968.

Beluga hunting is now limited to local residents. Much the same now applies to narwhals. Neither species is endangered, but concern is still needed to ensure, particularly in the case of the beluga, that local populations are not exterminated. Careful monitoring of catches, making full allowance for those which sink and are lost when killed, is essential if populations are to be maintained, let alone recover from earlier depredations.

The Irrawaddy dolphin

For many years it was believed that the beluga and the narwhal were the only two species in their family, the Monodontidae. Now, however, it is considered that a small tropical dolphin is more nearly allied to these whales than to either the killer whale group or the other dolphins. The Irrawaddy dolphin, *Orcaella brevirostris*, vaguely resembles a beluga, although it is smaller and has a prominent dorsal fin (Fig. 8.3). Fully grown, it is 2–2.5 m (6 ft 11 in–8 ft 2 in) long and weighs about 100 kg (220 lb). It is evenly blue-black on its back and flanks and paler beneath. The stout body has a rounded head with a distinct melon, but not even a

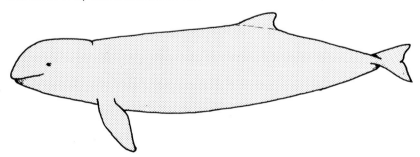

Fig. 8.3   The Irrawaddy dolphin, *Orcaella brevirostris*.

trace of the incipient beak seen in the beluga. There are 12–19 pairs of conical teeth in the upper jaw and 12–15 pairs in the lower. It is a coastal dolphin of the Indo-Pacific, being found from the Bay of Bengal through India, Pakistan, Burma, Vietnam and New Guinea, to northern Australia. It frequently enters the estuaries of the larger river systems, hence its vernacular name.

Little is known of this dolphin, apart from skulls preserved in museum collections. Its population size is quite unknown, but it is not rare. It tends to occur in small groups of up to ten. It is said to feed on the bottom in shallow muddy water, taking fish and crustaceans. Certain native fishermen claim that the dolphins co-operate with them by driving fish into their nets, and there have been court cases between rival fishermen, one charging the another with having lured away the plaintiff's dolphin. There is no fishery for these dolphins, but, as in so many other cases, there are many instances of them being trapped and drowned in gill nets.

Chapter 9
Beaked Whales and River Dolphins

Beaked whales

The beaked whales, members of the family Ziphiidae, are the least known of all the whales. With a few exceptions, such as the bottlenose whales, they have not been the subject of commercial fisheries; they are inconspicuous, surfacing only briefly without producing an obvious blow; and they do not occur in large schools or ride the bow wave of vessels. Most of what we know has been garnered from stranded specimens found on beaches. Some species are known only from a few skulls and there may yet be undescribed forms swimming in the deep oceans.

Beaked whales (Fig. 9.1) are medium-sized whales between about 4 and 12 m (12–40 ft). There is a small, sickle-shaped dorsal fin, set rather far back on the body, and the tail flukes, unlike those of all other cetaceans, have no central notch. As the name of the group implies, there is usually a distinct beak present, although the head shape varies between species. The jaws are always elongated, and seem rather fragile for the size of the animal concerned. Beneath the chin there are two characteristic diverging longitudinal grooves. Perhaps their most unusual character however, is their dentition. In all but two cases, teeth are entirely absent from the upper jaw and reduced to only two pairs in the lower jaw. In females of most species the teeth never erupt. The true function of these teeth is very puzzling.

Because of the obscurity of these whales, they tend to have no vernacular names in common use. The names I shall use here are those generally accepted by whale biologists, but they are used scarcely more often than the scientific Latin names.

Baird's beaked whale, *Berardius bairdii*, is the largest of the group. The females are slightly larger than the males, a very large female Baird's whale reaching as much as 12.8 m (42ft) and weighing about 15 tonnes. The head has the pronounced beak characteristic of the group, with an undershot lower jaw. In the tip of this are two pairs of large teeth, the front pair protruding beyond the tip of the upper jaw. The body is a uniform slaty-grey, with sometimes a few white patches on the belly and usually with many long scars, often in parallel pairs.

These are whales of the deep waters of the North Pacific, although they do occur in coastal waters off Japan, where they are hunted from the Chiba Prefecture, mainly for the sake of their meat, which, when dried, is greatly relished. They extend northward up into the Bering Sea and the Gulf of Alaska; their southern limit is about the latitude of Baja California.

These whales mate mostly in October and November and give birth in

156

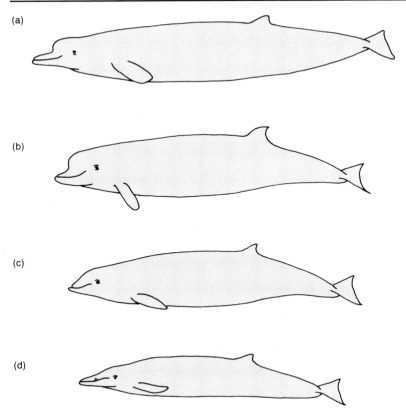

(a)

(b)

(c)

(d)

Fig. 9.1 Beaked whales: (a) Baird's, *Berardius bairdii*; (b) northern bottlenose, *Hyperoodon ampullatus*; (c) Cuvier's, *Ziphius cavirostris*; (d) Sowerby's, *Mesoplodon bidens*.

March and April of the next year but one, after a very long gestation period of 17 months. They reach sexual maturity at about eight years and may live for 70. Squid form a large part of their diet, but rays, deepwater fish and octopus are also taken.

A near relative of Baird's beaked whale is Arnoux's beaked whale, *Berardius arnuxii*. This is a smaller species, growing up to 9.8 m (32 ft 2 in), although very few have been measured. This whale has a circumpolar distribution in cold and cool waters around Antarctica, but reaching as far north as central Chile, Argentina and South Africa. Arnoux's whale, of which only about 30 specimens are known, is presumed to be similar in habits to Baird's whale, the two populations perhaps having separated in the last Ice Age.

The role of Baird's whale in the temperate and arctic North Atlantic is taken by the northern bottlenose whale, *Hyperoodon ampullatus* (Fig. 9.1). This is a smaller animal than Baird's whale and the females are smaller than the males, about 8.5 and 9.5 m (28 and 31 ft), respectively. As in

Baird's whale, there are a pair of protruding teeth at the tip of the lower jaw. Because these do not occlude against anything, they are a favourite site for the attachment of bunches of stalked barnacles and associated whale-lice. The barnacles are not very species-specific in their choice of hosts, but the whale-lice are, and it is interesting that the northern bottlenose whale and Baird's whale have lice of the same genus, *Platycyamus*, indicating that the whales already carried the lice when their ancestral stock divided.

The northern bottlenose whale is a migratory species. In the summer it moves north, even as far as the Arctic ice-edge, but in the winter it retreats to warmer waters as far south as the Cape Verde Islands. Most of the bottlenose whale strandings on the coasts of Europe occur in late summer and autumn, on their southward migration. Normally, however, this whale is found in waters with depths in excess of 1,000 m (3,300 ft). They are deep divers, and possibly go deeper even than the sperm whale. Diving times of up to two hours have been recorded.

Northern bottlenose whales have been hunted from Norway for over a hundred years. Unlike most beaked whales, they are inquisitive animals and will approach ships. They are also said to be extremely protective of their group members, so that if one animal were to be wounded, the others would gather round, making themselves easy prey for the whalers. Norwegian bottlenose whaling ceased in 1973; annual catches had averaged 381 for the decade 1962–71. In the 1890s peak annual catches had been in the region of two to three thousand, and it would seem that this stock has been reduced by commercial hunting.

A southern bottlenose whale, *Hyperoodon planifrons*, at the ice-edge in the Weddell Sea. (In fact, this might be a specimen of Arnoux's beaked whale, *Berardius arnuxii*. It is impossible to tell with certainty without detailed examination of the teeth.)

Cuvier's beaked whale, *Ziphius cavirostris*, stranded on a Californian beach. The pale snout is characteristic of this species.

The southern bottlenose whale, *H. planifrons*, is a very closely related species, found in cold and temperate waters of all the southern oceans. It is in all known respects very closely similar to its northern relation and has probably been separate from it only since the last Ice Age, like Baird's and Arnoux's.

Cuvier's beaked whale, sometimes known as the goose-beaked whale, *Ziphius cavirostris* (Fig. 9.1), is a cosmopolitan species, found in all oceans except those at high latitudes. Males seem to be a little smaller than the females, which reach about 7 m (23 ft), but this information is based on very few measurements. In the males there are a single pair of conical teeth at the tip of the lower jaw. The coloration is very variable; yellow to dark brown in the Pacific and grey to blue-grey in the Atlantic. There are often white oval scars on the body, and in adult animals the head and nape of the neck become white. The beak is less marked in this species and the curve of the jaw gives it a strong resemblance to the beak of a goose. Very little is known about this species, although a few are taken by the Japanese. The records we have of stomach contents indicate that this whale, like the four already discussed, feeds primarily on squid but also takes fish.

The next species, the Tasman beaked whale, *Tasmacetus shepherdi*, is even less known that Cuvier's whale. Ten beached specimens have been found in the Southern Hemisphere, in New Zealand, South Australia, Argentina

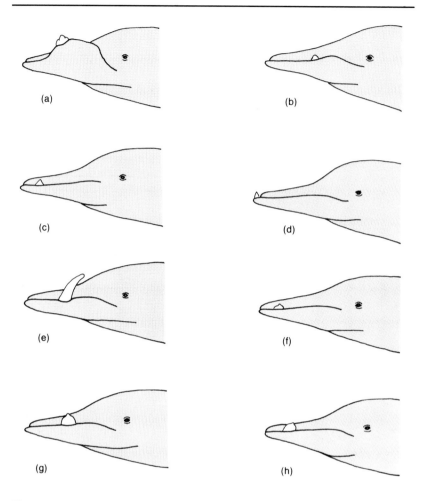

Fig. 9.2 Whales of the genus *Mesoplodon* differ mainly in the arrangement and shape of their teeth. (a) Blainville's beaked whale, *Mesoplodon densirostris*; (b) Sowerby's *M. bidens*; (c) Gervais', *M. europaeus*; (d) True's, *M. mirus*; (e) strap-toothed, *M. layardii*; (f) Gray's, *M. grayi*; (g) Stejneger's, *M. stejnegeri*; (h) Hubbs's, *M. carlhubbsi*.

and Chile. It probably reaches 6–7 m (20–23 ft) in length. Uniquely amongst beaked whales, it has small, conical teeth in both jaws, numbering 17–29 pairs. Additionally, there are two much larger teeth at the tip of the lower jaw; these are unerupted in females. No certain sightings of live Tasman beaked whales have been recorded and we know nothing of their natural history.

The remaining beaked whales are all grouped together in one genus, *Mesoplodon*. Some are moderately well known, others are known only from a couple of fragmentary skulls. The specific status of some of the forms may be uncertain and there could certainly be other forms still to be described. All are rather similar in appearance, but can be distinguished by their (sometimes extraordinary) teeth (Fig. 9.2). The body is spindle-shaped and generally somewhat laterally compressed, usually dark grey above and perhaps lighter below. Very often there are patterns of scars, sometimes parallel tramlines that might have been caused by the teeth of a con-specific. The forehead is not as bulbous or as bluff as in the first four beaked whales described. As in all the beaked whales, there are two throat grooves.

But it is to the teeth that we must turn if we wish to tell one *Mesoplodon* from another. This means, of course, that we have very little chance of telling the females or young of either sex apart, as in these the teeth usually fail to emerge from the gum.

A pair of massive teeth, each surmounted by a small nipple (the teeth themselves set on a huge, forward-pointing eminence on the lower jaw) marks Blainville's beaked whale, *M. densirostris*, perhaps the commonest of the genus, found throughout the Pacific, Atlantic and Indian Oceans. Flattened triangular teeth, near the middle of the beak, or at the end of the first third of the mandible, occur in Sowerby's beaked whale, *M. bidens*, mainly from the north-east Atlantic, but sometimes from the north west. Small triangular teeth, less than 10 cm (4 in) from the tip of the jaw, are found in Gervais' beaked whale, *M. europaeus*, from the warm temperate and subtropical Atlantic (although there is one specimen from the English Channel). True's beaked whale, *M. mirus*, from the North and South Atlantic and the south-west Indian Ocean, has small teeth, set leaning outwards, right at the tip of the jaw.

The teeth of the strap-toothed whale, *M. layardii*, are unmistakable.

The extraordinary jaws of Gray's beaked whale, *Mesoplodon grayi*. The solitary pair of huge, nippled teeth are situated half way along the lower jaw, but are so large that they protrude above the upper jaw, outside the mouth.

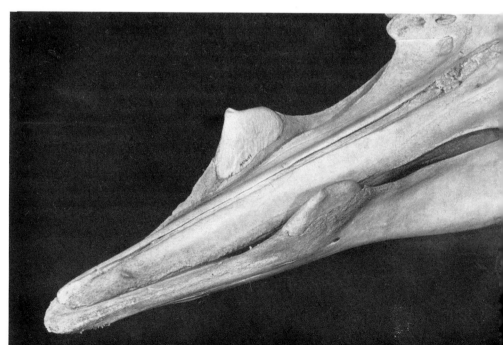

A rare sight. This Gray's beaked whale, *Mesoplodon grayi*, stranded itself on the flensing platform of a whaling station in South Georgia in the 1920s.

Emerging from near the tip of the jaw, a pair of huge, strap-like teeth slope backwards and over the beak, hindering the full opening of the mouth in old males. The strap-toothed whale has a circumpolar distribution in the temperate Southern Hemisphere, but only about 65 records are known. *Mesoplodon grayi*, Gray's whale, or the Scamperdown whale, has a pair of teeth with nipples on their tips, like those of Blainville's whale, but lighter in structure. Remarkably, this, alone among the *Mesoplodon* whales, has 17-22 pairs of tiny needle-like teeth in the upper jaw. Gray's is another temperate Southern Hemisphere whale, but, unaccountably, there is one stranding record from the Dutch coast!

A large pair of nippled teeth, splayed outwards and set at the posterior margin of the junction of the two halves of the lower jaw, identifies Andrews's beaked whale, *M. bowdoini*. The 11 known specimens are from New Zealand, Australia, Tasmania, Kerguelen – and Japan! *Mesoplodon pacificus* is known from two skulls only, one from Queensland and the other from Somalia. They both have very small teeth right at the tip of the jaw.

Hector's beaked whale, *M. hectori*, is better recorded, with 14 specimens, from Tasmania, New Zealand, South Africa, Tierra del Fuego, the Falklands and southern California. The Japanese, or ginkgo-toothed, beaked whale, *M. ginkgodens*, has a pair of enormous nippled teeth set on a bony boss at the end of the first third of the jaw. The teeth, which are much broader at the base within the socket, are said to resemble the leaf of the sacred ginkgo tree. The resemblance is lost on me. It is found in warm water from Sri Lanka, through the Japanese Sea to California. Stejneger's beaked whale, *M. stejnegeri*, is from the cold temperate and subarctic North Pacific. It has large nippled teeth, slightly forward-pointing, but set on a level jaw surface.

The last of this confusing genus is Hubbs's beaked whale, *M. carlhubbsi* (what a marvellously banal specific name, although Carl Hubbs was a marine biologist of note who certainly deserved commemorating). The teeth are similar to those of Andrews's beaked whale, but lean backwards, rather than forwards, and arch around the upper jaw, although in a less extreme manner than those of the strap-toothed whale. Hubbs's appears to be confined to the cold temperate North Pacific, although there are very few strandings (from Japan, Washington and California).

What can we make of these puzzling whales? The distribution patterns seem often to indicate a division into warm or cold-water species, or between Northern and Southern Hemispheres. But these are mostly based on very few specimens, and what principles we may try to derive are upset by the occurrence of 'rogue' specimens, like the Gray's beaked whale that turned up in the Netherlands, or the Andrews's beaked whale from Japan. Where there are indications of what food is taken, it seems to be basically squid, with occasional fish. Certainly the reduced dentition would seem to indicate a diet of soft-bodied squid. The teeth that remain, those diagnostic features, are almost certainly used in exactly this way by the whales themselves – to indicate membership of the right club. They are analogous to the colour patterns of many fish and birds, and even some mammals, the guenon monkeys of Africa, for example.

We know nothing about the genetic isolation of these various beaked whales, but it is difficult to believe that separation into the groups that we recognise today is anything other than very recent (in evolutionary terms). It would be exceedingly interesting to be able to capture some of these whales and keep them in captivity, to see what sort of hybridisation would take place. But perhaps the satisfaction of this intellectual curiosity would not justify snatching these mysterious creatures out of their proper environment for confinement in a concrete pool. We shall not be much the poorer for not knowing the precise relationship between *Mesoplodon carlhubbsi* and *Mesoplodon ginkgodens*!

No doubt some of the beaked whales are rare, and at least one species has been reduced by hunting, but none is threatened with speedy extinction. The next group of cetaceans to be considered, the river dolphins, is in a less happy position. Indeed, the only chance of survival for some of them may be to be artificially maintained in captivity.

River dolphins

River dolphins, or platanistids (Fig. 9.3), are among the most interesting of cetaceans. Although highly specialised for what is a rather untypical way of life for whales, they are generally accepted as the most primitive cetaceans living today. Their brains are small; only the Amazon dolphin, with a brain weighing 1.3 per cent of its body weight, compares with other dolphins. Their skulls do not show the extreme telescoping characteristic of other whales, and the various bones that make up the skull are more clearly delineated. There are numerous needle-like teeth in elongated and very slender jaws (between one-fifth and one-seventh of their overall body length). The seven cervical vertebrae are separate, giving the platanistids

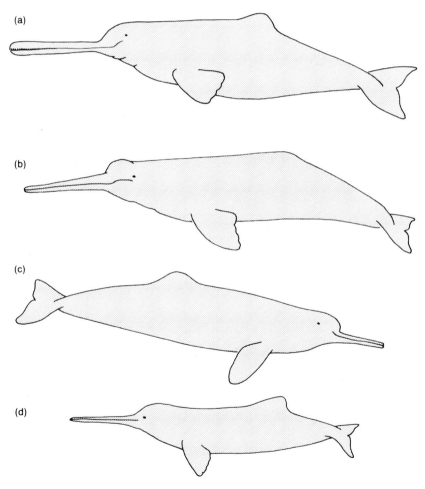

(a)

(b)

(c)

(d)

Fig. 9.3   River dolphins: (a) Ganges susu, *Platanista gangetica*; (b) boutu, *Inia geoffrensis*; (c) baiji, *Lipotes vexillifer*; (d) franciscana, *Pontoporia blainvillei*.

the ability, like the beluga, to turn their heads from side to side. Associated with this, there is a discernible neck. With one exception, the franciscana, they are confined to the freshwater of major river systems.

Two forms of river dolphin are found in the Indian subcontinent. These are usually described as separate species. *Platanista minor*, the Indus susu, is found, as the name suggests, in the Indus River system in the provinces of Sind and Punjab. *Platanista gangetica*, the Ganges susu, is found in the Ganges, Brahmaputra, Kharnaphuli and Meghna Rivers in India, Bangladesh and Nepal. There are negligible differences between these two forms (a matter of slightly lower bony crests on the skull of the Ganges susu), and it is likely that they are the same species, although their populations are genetically isolated.

The susus (an onomatopoeic name suggestive of the sighing noise they make when they exhale) grow to about 2–2.5 m (6 ft 6 in–8 ft 2 in) and weigh about 80–90 kg (176–198 lb). Each jaw is armed with 27–33 pairs of slender, sharp-pointed teeth. The blowhole is a longitudinal slit on the top of the head (in most other cetaceans it is transverse). The flippers are broad and paddle-like, with the outline of the digits showing clearly.

The susus' eyes are practically non-functional, as they lack a lens, but they can distinguish light from darkness. As the waters in which they live are thick with suspended silt, a sense of sight would be of little value to them, and it seems that, in vertebrates, if the eyes are not used, they tend to degenerate and become non-functional, as examples like blind cave fish and moles illustrate. Instead of sight, susus use echo-location to detect and capture their food. This is thought to consist mainly of fish and crustaceans. In hunting, the susu adopts a strange attitude, swimming along on its side, usually tilted to the left, and nodding continuously as it sweeps with its echo-location pulses.

Neither of the susus has been hunted for food, at least in historical times. In the sacred Ganges, at least, they are regarded with veneration. Relief from hunting pressures, however, does not ensure the survival of a species. The susus suffer from a more insidious threat – that of habitat destruction. The Indus River system has been greatly affected by the construction of dams and barrages built for irrigation purposes or hydroelectric schemes. These isolate the stocks of the dolphins and prevent them undertaking their usual migratory excursions associated with the monsoon – into the creeks and smaller branches during the rains, and a retreat to the main channels in the dry season. Siltation, brought about by restricting water flow, and by deforestation upstream, also reduces the suitable habitat. Most of the Indus susus are found in a 130-km (81-mile) stretch of the river between the Sukkur and Guddu barrages in Sind. Here, in 1979, there were no more than 290 dolphins, with very few elsewhere. Since then, however, the population has increased and, in 1986, there were thought to be between 400 and 600 animals, a rewarding result for the protection measures introduced by the Paskistan government.

The Ganges susu has so far been less affected by habitat destruction than its cousin, and perhaps 5,000 or so remain. Pollution is a major threat in the Ganges, but the susus seems to endure this remarkably well. Even in

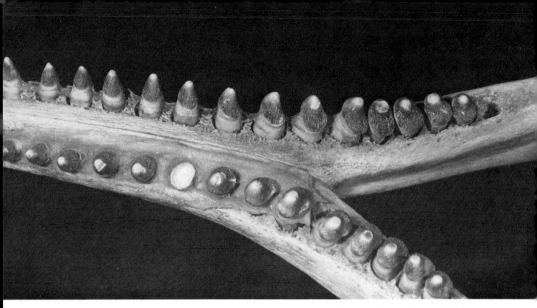

The crushing teeth at the back of the jaws of the boutu, *Inia geoffrensis*. This is about as far as modern whales go in a differentiated dentition, the last seven or eight pairs of teeth being more massive and having a shelf on their inner margins. Perhaps these are used for cracking the shells of fresh-water crabs.

the grossly polluted Hoogly River, around the industrialised area of Calcutta, the susus persist.

The boutu, *Inia geoffrensis*, is the river dolphin of the Amazon and Orinoco River systems. It is very similar in appearance to the susu, although perhaps slightly larger. Like the susu, it is pale greyish-brown, but occasionally has a pinkish tinge, particularly on the underside. The long, narrow beak of the boutu is armed with 24–34 pairs of teeth in each jaw. Exceptionally for a cetacean, these teeth are of two kinds; at the front of the jaws are simple conical teeth but, behind, the teeth are blunter and stouter and have been described as molariform. Another odd feature of this beak is the abundance of stiff, flattened sensory hairs that it bears. These are no doubt used for detecting food organisms in the muddy bottom of the rivers. The eyes of the boutu are reduced, but have lenses and are probably functional, as boutus, when surfacing, often bring the eyes clear of the water and appear to inspect their surroundings visually.

The food of the boutu is believed to consist largely of fish. In captivity they have been observed to take a fish with the teeth at the front of their jaws and then transfer it to the back, where it is chewed. It has been suggested that the molariform teeth may have been developed for crushing armoured fish, which abound in the tropical South American rivers. They would work equally well on crustacea and turtles, which are also abundant. During the rainy season, when the rivers burst their banks, boutus are observed deep in the forest, in the shallowest of the flood water, from which they may even slither out to reach a titbit on the mud.

Recent developments in the Amazon basin have menaced even this species, which is probably the least endangered of the river dolphins. Water impoundments for irrigation are probably not as much of a problem as in India, but hydroelectric schemes are, primarily because of the adverse effect of dams on the fish on which the dolphins feed. Commercial fishermen now operate along the length of the Amazon in Brazil and Peru. Not only do the dolphins become entangled in their nets and drown, but the fishermen consider them competitors to be shot whenever opportunity offers. As the boutu is a slow swimmer, and surfaces frequently, rarely staying down for more than a couple of minutes, these opportunities are frequent.

Some of the aboriginal tribesmen are said to believe that killing a boutu brings bad luck, but this has been more than offset by a bizarre trade that has sprung up in the more civilised(!) parts of Brazil. It seems that it is believed here that if a man possesses the dried eyeball of a boutu, or a woman its dried vulva, then sexual irresistibility is assured. The trade in these charms, which are available by mail order, is said to have spread from Brazilian towns to customers in France and Spain. There seems to be no end to the gullibility of men and women, but, in fairness, it should be said that there are no reliable statistics on this trade and Brazilian scientists have stated that it is of such limited extent that it is only a minor problem.

In the Yangtze River system lives the baiji (sometimes spelt beiji), or white-flag dolphin, *Lipotes vexillifer*. This is also very similar in appearance to the susu and the boutu, yet it seems these three river dolphins are not very closely related and we should attribute their similarity in appearance to the demands of the environments they inhabit – an example of convergent evolution. As in the susu and boutu, there is a low triangular ridge on the back – one can scarcely call it a dorsal fin – but in the baiji this is responsible for its specific name. '*Vexillifer*' means 'flag carrier' in Latin and was bestowed on the baiji because Miller, the original describer of this species in 1908, believed that the Chinese likened the fin of this dolphin to a flag. Baiji or '*pei c'hi*' actually means 'grey-white dolphin', a reference to the baiji's colour.

Until recently, little has been known in the Western world of the baiji, and the Chinese themselves have largely ignored it. Lately, however, there has been much more interest in this rare dolphin. In general, the Chinese

Yangtze river dolphin, or baiji, *Lipotes vexillifer*. This is Qi Qi, one of the specimens at the Institute of Hydrobiology, Wuhan, China.

fishermen had not molested the baiji, having a superstitious reverence for it. Inevitably, its oil was used for medicinal purposes, but there was no organised hunting. Unfortunately, traditional beliefs withered and casual exploitation for meat began. At the same time, more intensive fishing in the Yangtze led to many dolphins being caught on hooks or trapped in nets and drowned. Motor traffic on the river also resulted in dolphins being injured by collisions or being cut by propellors.

The Chinese authorities afforded the baiji full legal protection in 1975, but this did little to prevent the accidental mortality. In 1980 a young male baiji, caught on a fisherman's hook, was taken by scientists to the Institute of Hydrobiology in Wuhan and installed in a pool. Treated with Chinese medicine and salves, Qi Qi's wounds healed and he became healthy and active. The captive dolphin attracted a good deal of attention, to the benefit of the rest of this species. In 1984 two baiji reserves were set up on the Yangtze, totalling 255 km (158 miles) of river where fishing is banned and the banks are patrolled by dolphin guardians. There are plans to set up a reserve at Tongling. The site will consist of a channel of the river 12 km (7.5 miles) long between two islands. This would be netted off and the baijis fed from fish culture ponds. A research facility and a hospital for wounded animals would also be included.

Currently, there are three baijis, two males and one female, in captivity at Wuhan. Probably fewer than a thousand, and perhaps no more than a couple of hundred, now live in the Yangtze. Their best chance of survival as a species may be to live in semi-captivity in whatever part of the river the Chinese can spare them. The critical question is, as in the case of another Chinese rarity, the giant panda, will they breed successfully in captivity?

The last of the river dolphin group is not a river-dweller at all. The franciscana, *Pontoporia blainvillei*, lives in the coastal waters of western South America, from Peninsula Valdes in Chubut northwards to Ubatuba in Brazil. It is common in the estuary of the River Plate, and is sometimes known as the La Plata dolphin. It is the smallest of the platanistids, reaching no more than 1.7 m (5 ft 7 in), the females being slightly longer than the males. It has the same shaped head as the others, with the long slender jaws containing an astonishing number of teeth, 50–60 in each row! The blowhole is conventionally crescentic and there is a proper dorsal fin with a rounded peak. It is whitish-grey dorsally, and paler beneath. Young animals are brownish.

Franciscanas feed on fish, squid, octopus and shrimp, all obtained from the sea, for they never seem to enter the fresh waters of the rivers in the region. The franciscana is of scarcely any importance to fishermen, even as a competitor, but it suffers severely at their hands. Between 1960 and 1981 a minimum of 3,300 were drowned in gill nets, set chiefly for sharks, and it was suggested that off Punta del Diablo (certainly diabolical for the franciscanas!) in Uruguay the estimated annual catch was 2,000!

The four (or five) platanistids are fascinating little whales. It is sad that, having survived so long, we now seem to be edging them slowly out of the regions they still inhabit.

Chapter 10
Whales and the Modern World

For millennia whales have supported subsistence hunting by aboriginal peoples. For centuries some, like the right whales or sperm whales, have endured commercial exploitation. But it has been the last hundred years or so that have wrought the greatest changes for cetacea, from the river dolphins to blue whales.

In the 1860s a wealthy Norwegian sealing master began to experiment with a new method of hunting whales – but not the right whales which by then were already scarce, or the sperm whales which the Yankees pursued. Svend Foyn wished to be able to take the previously uncatchable rorquals, which, because of their speed and habit of sinking when dead, were all but ignored by the whalers of that time. (Whales at that time were hunted with a hand harpoon or shoulder gun used from an open boat, with a sailing vessel in attendance as a mother ship.) Foyn's concept was to mount, in the bows of a steam-driven vessel, a large cannon, firing a harpoon equipped with an explosive point and which carried a stout cable to fasten the struck whale securely to the vessel, so that it mattered not whether it floated or sank.

Guns of this sort already existed, but on a much smaller scale. Captain Roys, who opened up the bowhead fishery in the Bering Sea, had invented a rocket gun which was fitted with an explosive head, but this, like all its predecessors, was fired from an open boat. On its first trials, 52 whales were shot, but none saved. Equally, Foyn was far from successful at first, but his persistence and capital perfected a lethal device that was to deal a death blow to hundreds of thousands of the largest whales.

The first whale ship, evocatively named *Spes et Fides* (Hope and Faith), was not an instant success, but Foyn persevered and was soon reaping a profitable harvest in north Norway. By 1880 there were eight companies operating there with 12 steam catchers. The first foreign station was set up in Iceland in 1883 and it was no more than a dozen years later that there was one as far away as Japan. Local stocks of whales were rapidly reduced, but isolated whaling stations continued to operate around the shores of both the North Atlantic and North Pacific almost up to the present day, although nowhere on a very extensive scale.

It was in the Antarctic that the greatest war against the whales was to be waged. This industry was the brainchild of another Norwegian, Carl Anton Larsen. Larsen had accompanied the Swedish explorer, Otto Nordenskjöld, on a voyage to the Antarctic in 1901–1903. Here he was greatly impressed with the staggering abundance of large rorquals, blues, fins and humpbacks. Seeing these in terms of the oil they might produce, a very natural attitude at that time, Larsen raised capital in Buenos Aires

The Antarctic whaling industry did irrevocable harm to the stocks of baleen whales in the Antarctic. Here a 24.4-m (80-ft) fin whale, *Balaenoptera physalus*, lies on the flensing platform in South Georgia.

for a whaling expedition, and on 16 November 1904 he arrived in the harbour of Grytviken in South Georgia.

The first year's operation (when they were still building the station!) resulted in a catch of 183 whales, mostly humpbacks, it scarcely being necessary to go outside the limits of the bay in which Grytviken lay to find whales. The following season was even more successful, with 399 whales. The pattern was set. Other companies soon joined Larsen and by 1906 there were whaling companies established in the South Shetland Islands as well as South Georgia.

A few of these companies had shore stations, but most worked from floating factories, old steamers fitted with boilers and moored in some secure anchorage where there was a supply of fresh water and shelter to flense the whales alongside in the water. It so happened that suitable sites for shore stations and harbours for floating factories were all in that area of the Antarctic which was claimed, and administered after a fashion, by the United Kingdom through the Falkland Islands. At that time there was a very remarkable man, Sir William Allardyce, serving as governor, who was something of a conservationist for the time.

Foreseeing the possible, indeed, likely, destruction of the whale stocks, as had happened in the north, Allardyce limited the number of licences granted to operate whale catchers and the number of leases for factory sites. He made requirements to utilise the carcases as well as the blubber, and provided protection for young whales and whales accompanied by calves. It was not an ideal system, but it was in its way effective. Because control was in the hands of one powerful authority, the whaling companies toed the line.

All might have continued well had it not been that in the whaling season of 1912–13 a Norwegian floating factory, bound for its berth in the South Orkney Islands, and unable to get in because of the pack-ice, killed a few whales along the ice-edge, flensing them at the side of the ship at sea in the comparative shelter of the ice. This event sowed the seeds of another fatal idea for whales in the mind of one of the whale gunners aboard the *Tioga*.

Petter Sørlle reckoned that if a whale could be brought aboard a ship, it could be flensed up and cooked out on the high seas, without the troublesome need for a licence from the governor of the Falkland Islands, and, even better, without any liability to pay duty on the oil produced. Petter Sørlle's invention was the stern-slip factory ship, a large steamer with a slipway set in its stern, up which the whales could be hauled on to the flensing deck. The first such ship, the *Lancing*, sailed from Norway in June 1925. On her way south she took 294 humpbacks from their breeding grounds off the Congo before going to the Antarctic where she continued the slaughter.

In the absence of any centralised control, development was very rapid. In 1925–26 there were two expeditions operating pelagically, which together produced 56,814 barrels (9,652 tonnes) of oil. By 1930–31 there were 41 factory ships, which killed 37,500 whales to produce 3,420,410 barrels (579,120 tonnes) of oil. Such vast production glutted the market

Ripping the blubber from the carcase of a fin whale.

A fin whale, *Balaenoptera physalus*, being flensed alongside a floating factory in a sheltered cove at South Georgia in the 1920s.

The carnage on the flensing plan of a whaling station. An oily mass of blubber and bloody guts litters the deck while the workmen trudge through the mess selecting the next pieces to go into the boilers. South Georgia 1953.

and forced down the price of oil, so the first respite for the whales came from the whalers themselves, who agreed to lay off for a year and to introduce some control measures. It is just possible that, had the industry remained in Norwegian and British hands, some lessons might have been learnt from this. But this was not to be. In 1934 Japan entered Antarctic whaling, to be followed in 1936 by Germany. By 1939 there were six different flags flying above 34 factory ships in the Antarctic.

World War 2 provided a brief respite for the whales, but no sooner had the war finished than the world shortage of edible oils led to a speedy resumption of whaling. By 1950-51 there were 20 factory ships, which killed 32,566 whales. By now, however, there was a recognised contolling authority. This was the International Whaling Commission – the IWC. This was set up in 1946 and was responsible for the regulation of whaling, not, as some people imagined, the conservation of whales, although, of course, the two subjects were intimately linked.

The regulations introduced by the IWC provided absolute protection for grey whales and right whales, and for humpbacks in the Antarctic; certain areas of the Antarctic were to be out of bounds for whaling, which was to be confined to a specified season; no whales below designated minimum lengths were to be taken; shooting of whales accompanied by calves was forbidden; and, finally, a catch quota of 16,000 blue whale units was set.

A right whale and whaling gear ashore at Spitzbergen.

The blue whale unit (BWU) was an odd concept. One BWU was equivalent to one blue whale, two fin whales, two and a half humpbacks or six sei whales. Each of these combinations was equivalent in oil production to one blue whale.

This quota was probably not far above what could be taken on a sustained basis from the Antarctic whales, but, because it took no account of the varying degrees of protection required by the different stocks, it allowed a depleted stock to be further reduced while the whalers concentrated on a more abundant one.

The sad tale of Antarctic whaling is all too familiar. Attempts to reduce the quota in the face of rapidly falling stocks were fiercely resisted. At one stage, two important whaling nations, Norway and the Netherlands, withdrew from the convention rather than agree to a quota reduction. The quota came down to 15,000 BWU in 1962–63 (though only 11,306 BWU could be found by the whalers!), and to 10,000 BWU in 1963–64.

It was not until after further quota reductions, although still to levels not suffiently low to help the beleaguered whales, that the BWU was abolished in 1972–73. In 1975 the IWC introduced a new management policy. This looked at different stocks of whales and calculated quotas for each of them, in an attempt to provide protection where it was needed (almost everywhere!), but also a harvest where it could be taken on a sustainable basis.

By this time, however, it was all but too late for both the whales and the whalers. Although substantial stocks of minke whales remained in the Antarctic, which were exploited by the USSR and Japan, all other countries had dropped out of whaling on economic grounds. Calls were being heard from developed countries for a complete cessation of this trade, and the IWC, from being a group of whaling countries, was becoming dominated by those with protectionist motives. Following repeated calls for a moratorium on whaling, this was finally adopted by the IWC in 1982, to come into effect in 1986. Objections were made to this recommendation by several whaling nations, although the USSR and Japan undertook to end commercial whaling in the Antarctic in 1987. In fact, Japan has continued to whale in the 1987-88 Antarctic season, but for scientific samples, rather than a commercial catch. The Japanese justification for this approach (which has also been put forward by Iceland) is that in order to discover whether a stock of minke whales is in a state that would allow a sustainable harvest to be taken, it is necessary to sample the stock. This may be true, but the distinction between a scientific quota and a commercial catch will be obscure to a whale with a harpoon in its back.

Antarctic whaling was a bloody business. I spent ten years living next to a whaling station in the Antarctic. I grew accustomed, but never indifferent, to their great carcases lying on the flensing deck as, amid the roar of the steam winches, their oleaginous blubber was ripped off. The

Here a bottlenose dolphin is helped out of a tuna purse seine. Despite such efforts the mortality of dolphins in the tuna purse seine fishery is still unacceptably high.

method of killing – the internal explosion of a cast-iron grenade packed with 400 g (14 oz) of black powder – was undeniably cruel (although most of the whales died quickly). The philosophy of the industry was deplorable. The whalers were aware that they were destroying their own livelihood, but for the most part they had no other trade to turn to. Many governments, particularly those of Japan, the Netherlands, Norway, the UK and the USSR, failed to provide a strong policy lead to direct investment away from the whaling industry, and hence contributed as much to the destruction of the whales as did the whalers.

Whether or not commercial whaling will ever return to the Antarctic is debatable. Minke whales, hunted for meat to sell at premium prices in Japanese markets, might just be economical. But I doubt if the great whales will ever return in their former numbers. Their place in the ecosystem has been usurped by other krill-consumers – penguins and seals – whose numbers have increased dramatically since the whaling era.

Whale killing is not confined to the great whales, or even to commercial killing. We have seen the aboriginal take of bowheads and the drive fisheries for pilot whales in the Faeroes. At least the products from these whales are utilised. It is disheartening to know that unintentional massive killing of dolphins goes on without malice to the dolphins or benefit to the killers.

Tuna are much sought after by fishermen in the warmer waters all over the world. In the north-east tropical Pacific, yellow-fin tuna are commonly found in association with schools of dolphins. The reason for this is not known, but it may be associated with the dolphins' superior prey-locating techniques, using their sonar, which would be of benefit to the tuna, while these might, in turn, alert the dolphins to approaching sharks. Whatever the relationship, the association is a consistent one. Fishermen cruise the ocean on the look out for schools of dolphins, visible at long range by their leaping and generally boisterous activity. Once a school has been located the fishing boat sets out to encircle it, using a purse seine net some 1.5 km (nearly a mile) long. When the dolphins and whatever tuna might have been associated with them, are within the net, the bottom of the net is drawn together, or pursed, and the net gradually drawn in. The dolphins, in their panic, become entangled in the net and many may drown.

Since 1959 nearly seven million dolphins, mostly spinner, spotted, common and striped dolphins, have been killed in tuna fishing operations. This vast slarghter provided much of the impetus for the United States Marine Mammal Protection Act, which was passed in 1972. Regulations made under this Act forced American fishermen to take some precautions to lessen the mortality, by inserting panels of special mesh into their nets, which the dolphins could detect and over which they could escape. Foreign fishermen were required to demonstrate that they took similar precautions if they were to import their fish into the US, although regulations to enforce this have not been forthcoming.

For a time conditions improved, but there was soon a reversal of this trend. Between 1981 and 1986, dolphin mortality more than quadrupled.

American tuna fishing performance deteriorated and the foreign fleet, largely Mexican, increased. About 130,000 dolphins were killed in the tuna fishery in 1986, and their carcases thrown back into the sea. The American Tunaboat Association, concerned about the mortality (and possibly also with their own image) now makes a Lou Briot Golden Porpoise Award for successful porpoise release. In 1987 Captain Roman Rebelo was the winner. He had encircled over 137,000 dolphins in capturing 4,677 tonnes of tuna and only 184 dolphins had perished, a successful release rate of 99.87 per cent . At the same time, the US tuna fleet had been granted a permit entitling them to kill 20,500 dolphins in that year, and by the end of it just under 14,000 deaths had been recorded.

The dolphin population in the eastern tropical Pacific is estimated at between four and eight million, and it has been calculated that about 2 per cent would have to be killed each year for there to be a permanent effect on the stocks. Since 130,000 is over 3 per cent of the lower stock estimate, there is cause for concern that a permanent effect may be starting to appear. For a dolphin entangled in a purse seine net the chances of the effect being permanent are, of course, much greater!

Entanglement of this sort is deliberate, even if the actual killing is inadvertent. Another much less easily quantifiable form of mortality arising from net entanglement has also recently been worrying marine biologists. When synthetic fibres were introduced to the fishing industry, they were hailed with delight. They were stronger than natural fibres, and hence could be used as thinner, and therefore less visible, yarns, and they were rot-proof and would last indefinitely. Such fibres were of particular value in gill nets, huge lengths of netting, suspended from floats at the surface of the sea, in which fish become entangled. A single boat will set many miles of such nets, and there are many boats.

In the North Pacific, where comparatively good data are available, it is estimated that 170,000 km (106,000 miles) of gill nets are in use each year and that in a single night, during the May to September squid-fishing season, some 14,400 km (nearly 9,000 miles) of net are set. The problem is that not only fish, but birds, seals and small whales can also become entangled and drown. It is hard to collect figures for this, but between 10,000 and 20,000 Dall's porpoises are believed to drown yearly in the Japanese salmon gill-net fishery. Perhaps 42,000 small cetaceans drown in gill nets off Sri Lanka each year. One could go on!

Fishermen often lose nets in a storm, or discard damaged nets by the traditional method of throwing them 'over the wall'. Such nets, made of indestructable plastic, become 'ghost nets', drifting through the oceans and continuing to entangle fish and the seals and dolphins that come to investigate the former. The mortality arising from ghost nets cannot be quantified, but there are indications that, for some species, it may be frighteningly large.

There are other, less generally serious, threats to cetaceans from fishermen. Japanese fishermen on the island of Iki recognise dolphins as serious competitors for the stocks of fish that both exploit. It seems only logical to the fishermen that, when opportunity offers, they should drive

This sort of contact with a dolphin does much to establish a link between child and cetacean, which may lead to a greater concern for these animals.

herds of dolphins ashore and slaughter them, not for the products that they can yield (though these are often used), but simply to destroy what they regard as pests. In this they are encouraged by the Nagasaki prefectural government, which pays a bounty for each dolphin killed.

In the narrow waters of Tierra del Fuego the kingcrab fishermen see dolphins not as pests, but a usable resource – in this case as bait for their pots. Probably not many dolphins are killed in this way, but some of the species so used are regarded as scarce, if not rare.

Pollution is another threat that cetaceans, like all dwellers in the seas, must face. The oceans are generally regarded as the ultimate dumping ground for waste products, either directly, as when wastes are deliberately dumped at sea, or indirectly, by the discharge of effluents through river estuaries. Oceanic cetaceans are not much affected by this. The vastness of the oceans ensures that they have great buffering power and biological processes can deal with most wastes if they are sufficiently dilute. Enclosed waters, such as the Baltic Sea, the Mediterranean or the St Lawrence estuary, are a different matter. Modern analytical methods permit the detection of exceedingly low levels of contaminants. The discovery of 0.1 part per million of DDT in a whale indicates that it has been exposed to

..

ironie that one of products a whales produces to help sustain its species survival – fat (which is then converted into oil by humans) should turn around to be possible become the destruction of their own species thru oil spills.

it seems very

pollution, but it is unlikely that it will be adversely affected by it. On the other hand, much higher levels are reported – 1,796–2,695 ppm DDT in bottlenose dolphins off California, for example – and we can be less sure of the innocuousness of such levels. Sublethal effects may cause a failure to breed (this has been shown in some seals) and this, of course, spells ultimate extinction for a stock as surely as acute poisoning.

Oil pollution usually has a conspicuous environmental impact, and increased development of subsea reserves, as well as the steady leaking or discharge of oil from ships at sea, has led to disfiguring blotches of oil on many coasts and the pitiful spectacle of oiled sea birds lying dead along the strand. The few observations that we have on the reaction of whales to oil slicks give no conclusive results. Bottlenose dolphins can detect oil films and will avoid them. On the other hand, grey whales have been seen swimming directly through oil slicks without showing any avoidance reaction. No cetacean mortality has yet been attributed to oil contamination.

The greatest threat to wildlife generally throughout the world is habitat destruction, depriving whole assemblies of species of their place on the world's surface. By and large, marine animals are less affected by this, and

A bottlenose dolphin, *Tursiops truncatus*, displays its teeth as it allows its chin to be rubbed by a spectator at San Diego's Sea World.

179

oceanic forms least of all. However, even whales, which are generally thought of as oceanic, may have a need for inshore areas at certain stages of their life cycle. Development of the lagoons of Baja California would certainly have a very adverse effect on grey whales, which need them for safe breeding areas. But not all cetaceans are oceanic, and many forms with an inshore distribution, such as the harbour porpoise or the bottlenose dolphin, may be finding their available range becoming restricted. Those most at risk are almost certainly the river dolphins.

Associated with habitat destruction is human disturbance. We know little of the effect of this on whales, but it has been suggested that the development of so many coastal resorts, while not actually destroying the marine habitat, may make it unsuitable for marine mammals. Manmade noise in the marine environment may upset the delicate discrimination of sound, so vital to the sonar-using odontocetes. Research vessels using explosive charges in the course of seismic surveys may cause serious injury to whales or other living creatures in the neighbourhood of the explosions. The increasing use of compressed air for the sounding signal, instead of high explosive, has improved conditions somewhat, but for some purposes, explosives are essential. Even the benign interest of whale watchers approaching whales too closely may have bad effects on the whales, although I think it is possible to exaggerate this.

Every year a number of dolphins are captured alive for exhibition in dolphinaria. The ethics of this are debatable. Many individuals and some wildlife societies are strongly opposed to taking animals whose natural habitat is the sea, and confining them in pools which, however large by human standards, must still be tiny in the scale of the ocean. There is a good deal of evidence that many dolphins suffer extreme stress, which may prove fatal, in the course of catching, shipment and confinement, even when all reasonable precautions are taken to avoid this.

There is no doubt that some of the conditions under which dolphins are kept are unacceptable under any standard. At others (and these tend to be the large, well-patronised insitutions, like San Diego's Sea World), environmental conditions and the health of the inmates are carefully monitored by a staff of experts, who, in many cases, are clearly devoted to their charges. Under such conditions, dolphins can live for long periods and even breed successfully, a criterion often acccepted as indicating satisfactory conditions in captivity.

I enjoy visiting a well-run dolphinarium, and I have learnt much of whales in such places. I do not much enjoy some of the more theatrical routines that the trainers set up. The sight of a young man, figged up in eighteenth-century costume as Paul Revere, travelling on a killer whale's belly from one side of a pool to another, strikes me as insulting to both man and whale. But I am perhaps taking an overly sophisticated view. It is clear that very many people do enjoy such sights, and, particularly, they enjoy the close presence of, or even physical contact with, the whales that such exhibitions provide.

The best of them also provide details of the biology of their charges. At Vancouver Aquarium there are no theatre or circus tricks. The trainer explains about cetacean anatomy and biology while her subject obligingly displays its tail or opens its blowhole to command. The audience is most enthusiastic and the dignity of the whale is not debased.

I believe that such displays, even the ones I personally disapprove of, serve the valuable purpose of promoting public interest in whales. Would the American Marine Mammal Act ever have become law if it had not been for a dolphin character on television, which was taken to the hearts of a nation's children? Had it not been for Flipper, would the mass destruction of dolphins in the tuna fishery have worried the American public any more than the slaughter of kangeroos for the pet-food trade? I am unhappy about confining an animal like a killer whale or a bottlenose dolphin in a pool for the rest of its life, but I feel that such 'gaol-birds', as one eminent American whale biologist calls them, are performing a valuable service for the rest of their kind – getting the public on the side of the whales.

A less popular side of cetaceans in captivity relates to research. The US Navy has spent vast sums on whale research over the last 40 years. Most of this research has been classified, but some has also been made freely available, and we know far more about cetacean biology that we otherwise would had it not been for these studies.

Much of the research has been concerned with training dolphins to locate objects under water. In October 1987, the US Navy confirmed that it had sent a team of dolphins to the Persian Gulf to hunt for mines, releasing a storm of protest from animal-rights activists. A spokesman for the Naval Ocean Systems Center at San Diego explained that the sonar system of the dolphin was superior to anything that the Navy could build, and that the Navy would continue to do whatever they could to protect the lives of American personnel, including using dolphins. So far as we know, no dolphins have been killed or injured on mine-hunting assignments, but there is a suspicion that such mortality might not be reported, anyway. Hunting mines is clearly a non-belligerent, defensive activity. Dolphins could, of course, equally well be trained to attach magnetic mines to hostile vessels.

Some people hold that the ethics of these policies are no different from those of other forms of warfare. This view is countered by those who feel that to involve animals in this way is an unjustifiable exploitation. Such views are irreconcilable, but one thing is certain; while humans continue to war among themselves, the rest of the natural world on this planet will suffer.

Is there hope for the whales? In Western Europe and North America there has been, in the last 40 years, a great upsurge in our feelings of responsibility to the environment we live in. The conservation movement has become a dominant force in many countries and even politicians cannot now ignore it. Awareness of whales and sympathy for them are greater than they have ever been before. Even with this, however, we see that much reckless destruction of whales continues. Although no cetacean species has become extinct in the last thousand years, it cannot be much more than a decade or so before this becomes no longer true. The baiji or the cochito, which will go first? Or will it be the Ganges susu? The loss of a rare species is a tragedy, but the loss of one which is abundant would be far worse. Can the Pacific dolphins continue to survive the purse-seine losses? We have no room for complacency in our dealing with the whales.

But I am inclined to be optimistic. I do not believe that large-scale commercial exploitation of the great whales will begin again. Economic considerations and public opinion would be against it in most developed countries and sanctions might preclude less-developed countries from following this course.

Should great whales become abundant once more, then it is possible that small-scale harvesting might begin, but I would expect this to be rationally controlled. Against the inexorable increase of the human species, which is threatening the whole of the terrestrial ecosystem for living space, whales are better protected than many species. When the fight for survival against chronic starvation begins in earnest, as it will in many of the world's poorer countries, maximising the production from the land, where the speediest returns can be found, will take priority over marine harvesting (except at the subsistence level). Whales, indeed, may well survive the human species.

A thoughtful biologist cannot look at the future with much hope. Perhaps the best option is to fear the worst but do all we can to ensure the continuing integrity of our environment. Whales and dolphins are a small, but very fascinating, part of this. A greater understanding of them must enrich human life and help us to appreciate better our place in the ecosystem of which all life is a part.

A Classification of the Order Cetacea

Sub-order ARCHAEOCETI (extinct)

Family Protocetidae
Family Basilosauridae

Sub-order ODONTOCETI (the toothed whales)

Super-Family SQUALODONTOIDEA (extinct)

Family Agorophiidae
Family Squalodontidea
Family Rhabdosteidae
Family Squalodelphidae

Super-Family PLATANISTOIDEA

Family Acrodelphidae (extinct)

Family Platanistidae
    Platanista minor — Indus susu
    Platanista gangetica — Ganges susu
Family Pontoporidae
    Pontoporia blainvillei — franciscana
Family Iniidae
    Inia geoffrensis — boutu
Family Lipotidae
    Lipotes vexillifer — baiji

Super-Family DELPHINOIDEA

Family Ketriodontidae (extinct)
Family Albireonidea (extinct)
Family Monodontidae
    Orcaella brevirostris — Irrawaddy dolphin
    Delphinapterus leucas — beluga
    Monodon monocerus — narwhal
Family Phocoenidae
    Australophoecaena dioptrica — spectacled porpoise
    Phocoenoides dalli — Dall's porpoise
    Phocoena phocoena — harbour porpoise
    Phocoena sinus — cochito
    Phocoena spinipinnis — Burmeister's porpoise
    Neophocaena phocaenoides — finless porpoise

Family Delphinidae

Steno bredanensis	rough-toothed dolphin
Sotalia fluviatilis	tucuxi
Sousa chinensis	Indo-Pacific humpbacked dolphin
Sousa teuszii	Atlantic humpbacked dolphin
Lagenorhynchus albirostris	white-beaked dolphin
Lagenorhynchus acutus	Atlantic white-sided dolphin
Lagenorhynchus obscurus	dusky dolphin
Lagenorhynchus obliquidens	Pacific white-sided dolphin
Lagenorhynchus cruciger	hourglass dolphin
Lagenorhynchus australis	Peale's dolphin
Lagenodelphis hosei	Fraser's dolphin
Delphinus delphis	common dolphin
Tursiops truncatus	bottlenose dolphin
Stenella species	spotted dolphin
(including *S. attenuata*, *S. plagiodon*, *S. frontalis*, *S. dubia*)	
Stenella coeruleoalba	striped dolphin
Stenella longirostris	long-snouted spinner dolphin
Stenella clymene	short-snouted spinner dolphin
Lissodelphis peronii	southern right whale dolphin
Lissodelphis borealis	northern right whale dolphin
Cephalorhynchus heavisidii	Heaviside's dolphin
Cephalorhynchus hectori	Hector's dolphin
Cephalorhynchus eutropia	black dolphin
Cephalorhynchus commersonii	Commerson's dolphin
Peponocephala electra	electra dolphin
Feresa attenuata	pygmy killer whale
Pseudorca crassidens	false killer whale
Orcinus orca	killer whale
Grampus griseus	Risso's dolphin
Globicephala melaena	long-finned pilot whale
Globicephala macrorhynchus	short-finned pilot whale

Super-Family ZIPHIOIDEA

Family Ziphiidae

Berardius bairdii	Baird's beaked whale
Berardius arnuxii	Arnoux's beaked whale

Hyperoodon ampullatus	northern bottlenose whale
Hyperoodon planifrons	southern bottlenose whale
Ziphius cavirostris	Cuvier's beaked whale
Tasmacetus sheperdi	Tasman beaked whale
Mesoplodon densirostris	Blainville's beaked whale
Mesoplodon bidens	Sowerby's beaked whale
Mesoplodon europaeus	Gervais' beaked whale
Mesoplodon mirus	True's beaked whale
Mesoplodon layardii	strap-toothed whale
Mesoplodon grayi	Gray's beaked whale
Mesoplodon bowdoini	Andrews's beaked whale
Mesoplodon pacificus	Hector's beaked whale
Mesoplodon ginkgodens	ginkgo-toothed beaked whale
Mesoplodon stejnegeri	Stejneger's beaked whale
Mesoplodon carlhubsi	Hubbs's beaked whale

Family Kogiidae
    Kogia breviceps — pygmy sperm whale
    Kogia simus — dwarf sperm whale
Family Physeteridae
    Physeter catodon — sperm whale

Sub-order MYSTICETI (the whalebone whales)

Family Cetotheridae (extinct)
Family Aetiocetidae (extinct)
Family Balaenidae
    Balaena mysticetus — Greenland right whale or bowhead whale
    Balaena glacialis — black right whale
Family Neobalaenidae
    Caperea marginata — pygmy right whale
Family Eschrictiidae
    Eschrictius robustus — grey whale
Family Balaenopteridae
    Megaptera novaeangliae — humpback whale
    Balaenoptera musculus — blue whale
    Balaenoptera physalus — fin whale
    Balaenoptera borealis — sei whale
    Balaenoptera edeni — Bryde's whale
    Balaenoptera acutorostrata — minke whale

Guide to Further Reading

Beale, Thomas (1839, reprinted 1973), *The Natural History of the Sperm Whale*, The Holland Press, London. (This is an excellent account of how things were and were seen from a whaler 150 years ago, written by the whaler's surgeon.)

Bonner, W. Nigel (1980), *Whales*, Blandford Press, Poole.

Burton, Robert (second edition, 1980), *The Life and Death of Whales*, Andre Deutsch, London. (A well-written and non-technical account.)

Dow, George Francis (1925, reprinted 1985), *Whale Ships and Whaling, a Pictorial History*, Dover Publications, New York. (Mainly pictures, but with a useful introduction and informative captions.)

Evans, Peter G. H. (1987), *The Natural History of Whales and Dolphins*, Christopher Helm, London/Facts On File, New York. (This is the most recent, and by far the best, of general natural histories. It contains an excellent bibliography.)

Gaskin, D. E. (1982), *The Ecology of Whales and Dolphins*, Heinemann, London. (An excellent book, full of original ideas, but written by a specialist at a fairly high level.)

Leatherwood, Stephen and Randall R. Reeves (1983), *The Sierra Club Handbook of Whales and Dolphins*, Sierra Club Books, San Francisco. (The best work for species accounts and identification.)

Mackintosh N. A. (1965), *The Stocks of Whales*, Fishing News Books, London. (An old, but still very valuable, account of the biology of the great whales and the dynamics of their exploitation.)

Matthews, Leonard Harrison (ed) (1968), *The Whale*, George Allen and Unwin, London. (A well-written and profusely illustrated account for the general reader.)

Matthews, L. Harrison (1978), *The Natural History of the Whale*, Weidenfeld and Nicolson, London. (A well-written account, still worth reading.)

Melville, Herman (1851, reprinted many times), *Moby Dick, or the Whale*, Richard Bentley, London. (A fine story, containing much about whales, as then known.)

Minasian, Stanley M., Kenneth C. Balcomb and Larry Foster (1984), *The World's Whales*, Smithsonian Books, Washington. (A species-by-species guide with stunning photographs.)

Scheffer, Victor B. (1969), *The Year of the Whale*, Charles Scribner's Sons, New York. (A fascinating account of a year in the life of a sperm whale, from the whale's point of view.)

Scoresby, William (1820, reprinted 1969), *An Account of the Arctic Regions*, Archibald Constable, London. (A good account of the right-whale fishery.)

Slijper, E. J. (second English edition 1975), *Whales*, Hutchinson, London. (A version, revised by Richard Harrison, of what was a seminal work when first published in 1958.)

Tønnessen, J. N. and Johnsen, A. O. (1982), *The History of Modern Whaling*, C. Hurst, London. (The definitive, but turgid, account of the subject.)

Watson, Lyall (1981), *Sea Guide to Whales of the World*, Hutchinson, London. (This sets out to provide the means to identify all cetaceans at sea, a task despaired of by most professional and experienced workers on whales. However, it contains a great deal of very useful information.)

Wynn, Lois King and Howard E. Wynn (1985), *Wings in the Sea – the Humpback Whale*, University Press of New England, Hanover. (A charming account of a single species, with plenty of good science.)

INDEX

Numbers in *italic* refer to black and white illustrations.
Numbers in **bold** refer to colour plates.

Aetiocetidae 25, 26, 185
Agorophiidae 25, 26, 183
Alaskan Eskimo Whaling Commission (AEWC) 64
Allardyce, Sir William 170
Amazon dolphin *see Inia geoffrensis*
Andrews beaked whale *see Mesoplodon bowdoini*
Archeoceti 22–5, *24, 25,* 183
Aristotle 9, 113
Arnoux's beaked whale *see Berardius arnuxii*
arterio-venous anastomoses 22
Atlantic humpbacked dolphin *see Sousa teuszii*
Atlantic white-sided dolphin *see Lagenorhynchus acutus*
Australophocaena dioptrica (Spectacled porpoise) *117, 118,* 183

baiji *see Lipotes vexillifer*
Baird's beaked whale *see Beradius bairdii*
Balaena glacialis (right whale or black right whale) 43, 48–54, *48, 49,* **50, 51,** *52,* 55–60, 173, *174,* 185
Balaena mysticetus (bowhead or Greenland right whale) 20, 54–66, *54, 61,* 174, 176, 185
Balaenoptera acutorostrata (minke whale) *30,* **34,** *35,* 36–8, *36,* **38,** 44, 47, 174, *175,* 176, 185
Balaenoptera borealis (sei whale) *30,* 34, 43, 44, 106, *136,* 174, 185
Balaenoptera edeni (Bryde's whale) *30,* 36, 44, 185
Balaenoptera musculus (blue whale) 11, 20, *28,* 29–30, *30,* 32–3, 37, 44, 45–7, *46,* 135, *136,* 169, 174, 185
Balaenoptera musculus brevicauda (pygmy blue whale) 32
Balaenoptera physalus (fin whale) 11, *30,* 33–4, *33,* 44, 45, *46,* 72, 108, 135, 169, **170, 171,** *172,* 174, 185
Balaenopteridae (rorquals) 25, 28–30, *30,* 32–47, *40,* 68, 95, 106, 135, 169, 185
baleen 27, 33, 34, 37, 38, **38,** *39,* 40–2, *40, 41,* 44, 48, 52, *52,* 54, 56, 58, 62, 67, 68, 77, 81, 82, 83
baleen whales *see* Mysticeti
barnacles 49, 69, **70,** 158
Basilosaurus 23–5, *24*
beaked whales 156–63, *157, 160,* 185
beak 116, 121, 122, 124, 125, 128, 130, 156, 166
Beaky 109, **110**
Beale, Thomas 66
behaviour 44, 52–4, 72–3, 75–6, 87, 88–9, *88,* 109, 133–7, 140–1, 148, 152, 158
beluga *see Delphinaptera leucas*
Berardius arnuxii (Arnoux's beaked whale) 157, **158,** 184

Berardius bairdii (Baird's beaked whale) 156–8, *157,* 184
black dolphin *see Cephalorhynchus eutropia*
black right whale *see Balaena glacialis*
Blainville's beaked whale *see Mesoplodon densirostris*
blowhole **23,** 24, **24,** 84, *85,* **86,** 91, 165, 168
blubber 20, 22, 45, 56, 58, 59, 171, *171*
blue whale *see Balaenoptera musculus*
Blue Whale Unit (BWU) 173–4
bonnet 49
bottlenose dolphin *see Tursiops truncatus*
bottlenose whales *see Hyperoodon*
boutu *see Inia geoffrensis*
bowhead *see Balaena mysticetus*
brain 38, 108, 114–15, 164
Borowski G.H. 68
Bryde's whale *see Balaenoptera edeni*
bubble netting 72–3
Burmeister's porpoise *see Phocoena spinipinnis*

callosities 49–51, *49*
Caperea marginata (pygmy right whale) 66–7, *66, 67,* 185
captivity 16, 82, **82,** 102–3, **135,** 137, 139, 145, 166, 168, **178,** 179, **179,** 180, 181
Case 89, *91,* 100
Cave, Professor A.J.E. 38
Cephalorhynchus commersonii (Commerson's dolphin) **128,** 129, *129,* 184
Cephalorhynchus eutropia (black dolphin) 128, 184
Cephalorhynchus heavisidii (Heaviside's dolphin) 128, *129,* 184
Cephalorhynchus hectori (Hector's dolphin) 12, *129,* 184
Cetotheridae 25, *25,* 26
chromatin bodies 76
clymene dolphin *see Stenella clymene*
Commerson's dolphin *see Cephalorhynchus commersonii*
common dolphin *see Delphinus delphis*
conservation 63–4, 65–6, 71, 80–1, 100, 154, 168, 173–5, 176–7
Cuvier's beaked whale *see Ziphius cavirostris*

Dall's porpoise *see Phocoenoides dalli*
Delphinus delphis (common dolphin) 11, 15, *15,* 108, 116, 120–1, *121,* 176, 184
Delphinapterus leucas (beluga or white whale) 55, 60, 146, 149–54, *149,* **150, 151,** 165, 183
diving 89–93, 95–7, 101, 158
dolphins 9, 11, 15, *15,* 19, 25, 102–46, *103, 104, 105, 107,* **110,** *112, 114,* **119,** *120, 121, 122,*

123, *124*, *125*, *126*, *127*, **128**, *129*, *130*, *137*, *139*, *140*, *141*, 176, 177, 178, 179, 180, 181, 182
Dorudon 25
drag 106, 107
dusky dolphin *see Lagenorhynchus obscurus*
dwarf sperm whale *see Kogia simus*

echolocation 113–15, *114*, 165, 180
Edge, Thomas 56
electra dolphin *see Peponcephala electra*
Enderby, Samuel 60, 98
energy budgets 44–7, *46*
entanglement 118, 121, 126, 155, 167, 168, **175**, 176–7, 182
epaxial muscles 105 *105*, 106
Eschrictius robustus (grey whale) 42, 76–83, *77*, *78*, **82**, 136, 173, 179, 180, 185
Euphausia superba 42–3, *43*, 45, 47, 52, 71, 72
evolution 9–10, 22–7, *25*, 30, 42, 83, 158, 159, 163, 167

false killer whale *see Pseudorca crassidens*
feeding and food 31–2, 37–47, *39*, *40*, *43*, *46*, 52, 55, 71–3, 81–3, 93–5, 101, 118, 126, 128, 129, 133, 134, *138*, 149, 152, 155, 157, 159, 163, 165, 166, 168
Feresa attenuata (pygmy killer whale) 138–9, *139*, 184
finless porpoise *see Neophocaena phocaenoides*
fin whale *see Balaenoptera physalus*
flick feeding 72
flipper 20–1
flukes 14, 20–1, 26, 102, 103–7, *104*
Foyn, Svend 169
franciscana dolphin *see Pontoporia blainvillei*
Fraser, Francis 112
Fraser's dolphin *see Lagenodelphis hosei*

Ganges susu *see Platanista gangetica*
Gaskin D.E. 44, 95
Gervais' beaked whale *see Mesoplodon europaeus*
geomagnetic contour 115–16
Gigi 82
ginkgo-toothed whale *see Mesoplodon ginkgodens*
Globicephala macrorhynchus (short-finned pilot whale) 15, *15*, 110, 141–3, *141*, 184
Globicephala melaena (long-finned pilot whale) 15, *15*, 110, 141–5, *141*, **142**, **143**, *145*, 176, 184
goose beaked whale *see Ziphius cavirostris*
Grampus griseus (Risso's dolphin) 140–1, *140*, 184
Gray's beaked whale *see Mesoplodon grayi*
Greenland right whale *see Balaena mysticetus*
Greenpeace 80
grey whale *see Eschrictius robustus*
Gudmundsson, Jon 78
gulping 42

habitat destruction 165, 166–7, 179–80
hair 16–17, 39–40, 49, 166
harbour porpoise *see Phocoena phocoena*
head 15, *15*, 20, 24, *39*, 84, 89, 90–1, *91*, 111
hearing 111–13, *112*
heat exchangers 20–2, *20*, *21*, 92
heat retention 17–20
Heaviside's dolphin *see Cephalorhynchus heavisidii*
Hector's beaked whale *see Mesoplodon hectori*

Hector's dolphin *see Cephalorhynchus hectori*
Hedges, Mitchell 29
hides 153, 154
hourglass dolphin *see Lagenorhynchus cruciger*
Hubb's beaked whale *see Mesoplodon carlhubbsi*
humpbacked dolphins *see Sousa*
humpback whale *see Megaptera novaeangliae*
Hussey, Captain Christopher 97
hypaxial muscles 105, *105*, 107
Hyperoodon ampullatus (northern bottlenosed whale) 95, 157–8, *157*, 185
Hyperoodon planifrons (southern bottlenosed whale) 95, **158**, 159, 185

Indo-pacific humpbacked dolphin *see Sousa chinensis*
Indus susu *see Platanista minor*
Inia geoffrensis (boutu or Amazon dolphin) *164*, 166–7, 183
insulation *see* heat retention
intelligence 107–10, 133, 136
International Whaling Commission (IWC) 63, 64, 65, 66, 80, 173–5
Irrawaddy dolphin *see Orcaella brevirostris*

Jacobson's organ 39
Japanese beaked whale *see Mesoplodon ginkgodens*
junk 89, *91*, 93, 100
Jurasz, Charles and Virginia 72

Kahn, Mukarrab 146
killer whale *see Orcinus orca*
Klinowska, Margaret 115, 116
Kogia breviceps (pygmy sperm whale) *100*, 101, 185
Kogia simus (dwarf sperm whale) **99**, *100*, 101, 184
Krill *see Euphausia superba*

Lagenodelphis hosei (Fraser's dolphin) *122*, 124, 184
Lagenorhynchus acutus (Atlantic white-sided dolphin) 122, *125*, 184
Lagenorhynchus albirostris (white beaked dolphin) 122, *122*, 184
Lagenorhynchus australis (Peale's dolphin) 124, 129, 184
Lagenorhynchus cruciger (hourglass dolphin) 122–4, *122*, 184
Lagenorhynchus obliquens (Pacific white-sided dolphin) 122, **123**, 184
Lagenorhynchus obscurus (dusky dolphin) 122, *124*, 184
laminar flow 106
La Platta dolphin *see Pontoporia blainvillei*
length 29, 32, 33, 34, 36, 37, 47, 48, 54, 67, 68, 77, 86, 89, 101, 117, 118, 119, 120, 121, 122, 124, 125, 128, 130, 137, 139, 140, 142, 147, 150, 154, 156, 157, 160, 165, 168
Larsen, C.A. 169
Lawrence, Barbara 150
Lipotes vexillifer (baiji or white flag dolphin or Yangtze river dolphin) *164*, 167–8, *167*, 181, 183
Lissodelphis borealis (northern right whale dolphin) 128, 184
Lissodelphis peronii (southern right whale dolphin)

127, 128, 129, 184
Lockyer, Christina 44–7
longfinned pilot whale *see Globicephala melaena*
long snouted spinner dolphin *see Stenella longirostris*
lunge feeding 72, 73

Mackintosh, N.A. 38
McTavish, Simpson 153
magnetite 16
Marine Mammal Protection Act (US) 176, 181
Matthews, L. Harrison 56, 108
meat 56, 63, 143
Megaptera novaeangliae (humpback whale) **23**, 27, 44, 68–76, *68*, *69*, **70**, **71**, 169, 170, 173, 174, 185
melon 114, *114*, 152, 154
melon-headed whale *see Peponocephala electra*
Melville, Herman 56, 84, 102
Mesoplodon bidens (Sowerby's beaked whale) *157*, *160*, 161, 185
Mesoplodon bowdoini (Andrew's beaked whale) 162, 163, 185
Mesoplodon carlhubbsi (Hubb's beaked whale) *160*, 163, 185
Mesoplodon densirostris (Blainville's beaked whale) *160*, 161, 162, 185
Mesoplodon europaeus (Gervais' beaked whale) *160*, 161, 184
Mesoplodon ginkgodens (Japanese beaked whale or ginkgo-toothed whale) 163, 185
Mesoplodon grayi (Gray's beaked whale) *160*, *161*, 162, *162*, 163, 185
Mesoplodon hectori (Hector's beaked whale) 163, 185
Mesoplodon layardii (strap-toothed whale) *160*, 161–2, 185
Mesoplodon mirus (True's beaked whale) *160*, 161, 185
Mesoplodon stejnegeri (Stejneger's beaked whale) *160*, 163, 185
migration 37, 45, 47, 51, 54, 61, *61*, 69–70, *69*, 79, 80, 81, 148, 153, 158, 165
minke whale *see Balaenoptera acutorostrata*
Monodon monoceros (narwhal) 146–9, *146*, **151**, 183
Moby Dick 84, 86, 94
Møhl, Bertil 115
Mysticeti (baleen whales or whalebone whales) *24*, 25, *25*, 26–7, 28–83, 112, 185
myoglobin 96

narwhal *see Monodon monoceros*
naso-frontal sac 91
navy (US) 74, 181
nekton 42
Neophocaena phocaeoides (finless porpoise) *117*, 119, 183
Nordhoff, Captain C. 73
Norris, Kenneth 114, 115
northern bottlenosed whale *see Hyperoodon ampullatus*
northern right whale dolphin *see Lissodolphis borealis*

Odontoceti (toothed whales) 24, *24*, 25–6, *25*, 31, 38, 84–165, 180, 183

oil 29, 45, 58, 60, 79, 97, 98, 100, 107, 135, 141, 143, 152, 153, 154, 168, 171, 173
orca *see Orcinus orca*
Orcaella brevirostris (Irrawaddy dolphin) 154–5, *154*, 183
Orcinus orca (killer whale or orca) 130–7, *130*, **131**, *132*, *134*, **135**, *136*, 180, 181, 184

Pacific white-sided dolphin *see Lagenorhynchus obliquidens*
Pakicetus 22, 23
Payne, Katy 74
Payne, Roger 49, 51, 74
Peale's dolphin *see Lagenorhynchus australis*
Pelorus Jack 140–1
Peponocephala electra (electra dolphin or melon-headed whale) 139–40, *139*, 184
periarterial venus plexus 20, *20*
Phocoena phocoena (harbour porpoise) 116–18, *117*, *118*, 180, 183
Phocoena spinipinnis (Burmeister's porpoise) *117*, 118, 129, 183
Phocoenoides dalli (Dall's porpoise) *117*, 118, 177, 183
Physeter catodon (sperm whale) **18**, 60, 64, 84–101, *84*, *85*, **86**, **87**, *88*, 91, 108, 130, 185
pilot whales *see Globicephala*
plankton 38, 42, 72
Platanista gangetica (Ganges susu) *164*, 165, 182, 183
Platanista minor (Indus susu) 165, 183
pollution 165, 178–9
Ponting, Herbert 133–4
Pontoporia blainvillei (franciscana or La Plata dolphin) 19, *164*, 165, 168, 183
porpoises 116–19, *117*, 183
Pseudorca crassidens (false killer whale) 137–8, *137*, **138**, 184
Purves, Peter 112, 114
pygmy blue whale *see Balaenoptera musclus brevicanda*
pygmy killer whale *see Feresa attenuata*
pygmy right whale *see Capera marginata*
pygmy sperm whale *see Kogia breviceps*

reproduction 53–4, 61, 76, 79, 87, 88–9, 118, 124–5, 149, 152
respiration 95–6
retia 21, 96, 97
Rice, Dale 84
right whale *see Balaena glacialis*
Risso's dolphin *see Grampus griseus*
Risting, Sigurd 29
river dolphins 25, *25*, 164–8, *164*, *166*, *167*, 168, 183
rorquals *see Balaenopteridae*
rough-toothed dolphin *see Steno bredanensis*
Roys, Captain Thomas 60, 169

Sandloegja 78, *78*
Scammon, Captain Charles 79, 80, 136
Scheffer, V. B. 27
Schevill, W. E. 44, 150
Schreiber, O. W. 74
Scoresby, Captain William 57, 58, 59
Scrag whale 78

Sea Life Park, Hawaii 139
Sea World, San Diego 16, 82, **179**, 180
sei whale *see Balaenoptera borealis*
sexual dimorphism 31, 86, 130, 142
short-finned pilot whale *see Globicephala macrorhynchus*
short snouted spinner dolphin *see Stenella clymene*
sight 85, 110–11, 165, 166
Silverman, Helen 148
size 19, 29–30, 47, 55
skeleton 12–14, *13*, 40, 84
skimming 42, 43, 52
skin 16, 22, 86, 92
skull 24, 25, 111, 112, 114, *118*, 119, 164
smell 38–9
song *see* vocalisation
Sørlle, Petter 171
Sotalia fluviatilis (tucuxi) 119, *120*, 184
Sound Fixing and Ranging (SOFAR) Station 74
Sousa chinensis (Indo-Pacific humpbacked whale) 119–20, **119**, *120*, 184
Sousa teuszii (Atlantic humpbacked dolphin) 119–20, 184
southern bottlenose whale *see Hyperoodon planifrons*
southern right whale dolphin *see Lissodelphis peronii*
Sowerby's beaked whale *see Mesoplodon bidens*
spectacled porpoise *see Australophocaena dioptrica*
spermaceti 90, 91, 97
spermaceti organ 89, 91–3, *91*, 95
sperm whale *see Physeter catodon*
spinner dolphins 109, 125–6, 176, 184
spotted dolphins *see Stenella sp.*
Squalodontoidea 25, *25*, 183
Stejneger's beaked whale *see Mesoplodon stejnegeri*
Stenella clymene (clymene dolphin or short snouted spinner dolphin) 109, 125–6, **176**, 184
Stenella coeruleoalba (striped dolphin) 125, *126*, 176, 184
Stenella longirostris (long snouted spinner dolphin) 109, 125–6, 176, 184
Stenella sp. (spotted dolphins) 126, *127*, **127**, 176, 184
Steno bredanensis (rough-toothed dolphin) 119, *120*, 184
streamlining 14–16, 106, 107
stranding 66, 101, 109, 110, 115–16, 138, 140, **142**, 143, 156, 158, **159**, 162, *162*, 163
strap toothed whale *see Mesoplodon layardii*
striped dolphin *see Stenella coeruleoalba*
suction feeding 82
swimming 14–15, 102–7, *103*, *104*, *105*, *107*, 141

Tasmacetus shepherdi (Tasman beaked whale) 159, 185
Tasman beaked whale *see Tasmacetus shepherdi*

taste 39
teeth 24, 25, 26, 41, 84–5, **87**, 101, 116, 117, 118, *118*, 119, 122, 124, *124*, 126, 128, 132, 137, 140, 142, 146, 147, 152, 155, 156, 158, 159, 160, *160*, 161, *161*, 162, 163, 164, 165, 166, *166*, 168
tongue 38, 39, *39*, 40, *40*, 41, 52, *52*, 82, 83, 94, 135
toothed whales *see* Odontoceti
touch 39–40, 166
throat grooves 34, 38, 40, *40*, 52, 69, 77, 83, 161
True's beaked whale *see Mesoplodon mirus*
tucuxi *see Sotalia fluviatilis*
turbulence 106, 107
Tursiops truncatus (bottlenose dolphin) 101, 108, 109, **110**, 113, 115, **123**, 124–5, *127*, 143, **175**, 179, **179**, 180, 181, 184
Tyack, Peter 76

Vancouver Aquarium 180–1
ventral grooves *see* throat grooves
vestibular sac 91, *91*
vibrissae *see* hair
vocalisation 73–6, 109, 113–15, 149–50, 153
vomero-nasal organ *see* Jacobson's organ

Washington Convention 100
Watkins, W.A. 44
water
    modification to 10–22, *10*, *11*, *13*, *15*, *19*, *20*, *21*
    properties of 12, 17–19, 111, 112
Watling, Frank 74
weight 19, 29, 30–1, 45, 47, 68, 86, 89, 139, 140, 148, 154, 165
whalebone *see* baleen
whaling 29, 45, 121, 146, 158, 168, 177, 178
    Antarctic 28–9, *28*, 70, **86**, **87**, *136*, *162*, 169–76, **170**, **171**, *172*, *173*
    bowhead 56–60, 61–6
    Eskimo 61–6, 145, 149, 153–4
    Faeroese 144–5
    grey whale 77–81
    Japanese 29–30, 145
    right whale 48, 55–60, **57**, **59**, 174
    sperm whale **86**, **87**, 97–100
    Stone Age 143–4
whale-lice 49, *50*, 69, **70**, 158
white beaked dolphin *see Lagenorhynchus albirostris*
white flag dolphin *see Lipotes vexillifer*
white-sided dolphins 122, **123**, 125, 184
white whale *see Delphinapterus leucas*
Wynn, Howard and Lois 74, 187

Yangtze river dolphin *see Lipotes vexillifer*

Ziphius cavirostris (Cuvier's beaked whale or goose beaked whale) *157*, 159, **159**, 185